The Science of Putting

*An analysis of the Personal Pendulum Putting Tempo, Ball Roll Physics,
Backstroke Length, Biomechanical Putting Posture and
the Physics of Moment of Inertia in Putting*

Steven R. Davis
and
Michael J. Davis

Copyright © 2021, Steven R. Davis and Michael J. Davis
All rights reserved.
Printed in the United States of America.
Except as permitted under the United States Copyright Act of 1976,
no part of this publication may be reproduced or distributed in any form
or by any means, or stored in a database or retrieval system,
without the prior written permission of the publisher.

The Science of Putting
An analysis of the Personal Pendulum Putting Tempo,
the Biomechanical Posture and Arc of the Putting Stroke ,
and the Physics of Moment of Inertia in the Putting

ISBN: 978-0-9793038-1-4

Published by
Painters Hill Press
Flagler Beach, FL

Cover design by
Pixelworks
John Lashbrook
Pompano Beach, FL

This publication is designed to provide accurate and authoritative information in regard to the subject matter covered. It is sold with the understanding that neither the author nor the publisher is engaged in rendering legal, engineering or other professional service.

White Paper

by the CURE Putters people

Table of Contents   Foreword; Acknowledgements; Introduction

Section 1   **Pendulum Putting Tempo**
    1.1   The Modern Pendulum Tempo Putting Stroke
    1.2   Theoretical Pendulum Physics
    1.3   The Range of Pendulum Putting Tempos
    1.4   Constant Acceleration is Key
    1.5   Practicing the Pendulum Putting Tempo

Section 2   **Ball Roll Physics**
    2.1   Ball Roll Physics - Launch, Skid and Roll
    2.2   Stimpmeter Physics – Bounce and Roll
    2.3   Impact Ball Speed and Putt Distance
    2.4   Impact Ratio

Section 3   **Backstroke Length Physics**
    3.1   Putter Speed and Backstroke Length
    3.2   Putter mph per Backstroke Inch
    3.3   Backstroke Length Controls Putt Distance

Section 4   **Sloping Greens Backstroke Length Physics**
    4.1   Effective Stimps for Sloping Greens
    4.2   Backstroke Length Adjustment for Sloping Greens
    4.3   Backstroke Length Card for Sloping Green
    4.4   Why Stimps are Exaggerated

Section 5   **Biomechanical Putting Posture**
    5.1   The Two-Plane Neutral Arc Posture Dominates the Tour
    5.2   The Elliptical Arc of the Putting Stroke
    5.3   Square to the Arc Face Rotation Varies with Posture
    5.4   Rotational Stability in the Putting Stroke
    5.5   Biomechanical Rotation on the Putting Stroke Axis
    5.6   Biomechanical Fitting Protocol
    5.7   The Neutral Putting Arc Face Rotation
    5.8   Measuring Spine Angle Accurately is Important

Section 6   **Moment of Inertia (MOI) and Stability**
    6.1   Putter Stability vs Putting Stroke Stability
    6.2   Moment of Inertia is Stability, Measured
    6.3   Impact Stability (Forgiveness)
    6.4   Weight Optimization Fitting and Pendulum Physics
    6.5   Stroke Stabilization (Face Angle Variation Reduction)
    6.6   MOI Summary of Findings
    6.7   Weight Optimization Fitting Protocol

Section 7   **Green Speeds, The Modern Putting Stroke and the Modern Putter**
    7.1   A Short History of Green Speeds
    7.2   The Evolution of the Modern Putting Stroke
    7.3   The Evolution of the Modern Putter

Section 8   **Putting Myths and Misunderstandings**
    8.1   Putting Myths and Misunderstandings

Section 9   **Putting Launch Monitor Metrics**
    9.1   Putting Launch Monitor Metrics

Section 10   **Miscellaneous Putting Fundamentals**
    10.1   Alignment, Grip and Stance

Section 11   **Miscellaneous Technical Information**
    11.1   Excess Exit Angle from Gear Effect Spin
    11.2   Overview of Historical Putter MOI
    11.3   Moment of Inertia Definitions
    11.4   The Effect of Shaft Stability on Putter Stability
    11.5   The Dimple Effect

Appendices   **A Backstroke Length Cards   B Formulas**

Foreword

Rod Spittle

Champions Tour Winner, Two-time Canadian Amateur Champion, member of the Canadian and Ontario golf Halls of Fame, Ohio State golf alum and member of the Ohio Golf Hall of Fame.
Rod is an inspirational teacher and respected instructional voice in the industry.

Greetings,

I am humbled and excited to introduce "The Science of Putting" to everyone. Steve Davis and his son, Michael, have written an outstanding book. The beauty of this book is that the reader will find that the majority of this collected data has been researched by people, not only fluent in Golf - but also Math and Physics, as well. Please note that I am not one of them! With that said, the ideas and concepts here have been dissected and illustrated in a manner that will help players of all skill levels enjoy and improve their golf games.

I grew up in Niagara Falls, Ontario, where like most Canadian youngsters, I had a hockey stick in my hands during the snowy months of winter. However, when summer grass eventually returned, my buddies and I lived at the Willo-Dell Golf Club. It was there that I chased my early dreams of playing golf like Arnold Palmer. So, in 1974, I received my Scholarship offer from Coach Jim Brown to play College Golf at The Ohio State University in Columbus, Ohio - my young career was off and running. My time at Ohio State was truly magnificent - with memories and friendships that I cherish to this day. However, the rest of the story is that I walked into the Ohio State Locker Room with a bag full of Persimmon woods and balata golf balls - yet, I had never seen my golf swing on tape or video. How about that - the technology just did not exist at that time. How times have changed!!

One of the luxuries of a Tour Professional is having access to the BEST equipment. In 2014, while on the PGA Tour Champions, I was fitted for a putter by Steve Davis. I learned that the weight-optimized, high MOI (Moment of Inertia) putter was a lot more stable; not just at impact but during the stroke itself. The putter simply stayed more square to the arc and twisted less at impact. That's Nirvana for all players and it's why the CURE putter is still in my bag!!

I now teach a seminar called "Think Like A Champion". One of the chapters deals with "unconventional" thinking - i.e. thinking Outside-the-Box. I have always believed that the BEST players do their "stuff" differently (eat, sleep, play, breathe, practice, read, study) because that is how they got to be the BEST. At the highest level, the BEST players chase Constant Improvement and Excellence.

It is for this exact reason that I accepted the invitation from Steve and Michael to open their book - because it pushes the boundaries of Conventional Thinking and Status Quo. I believe that "The Science of Putting" is on the cutting edge of the next wave of analysis and technology – and will revolutionize our amazing game.

As we all continue to chase Improvement and Excellence, I trust that you will enjoy the read.

Warm regards,

Rod

Acknowledgements

As father and son we have learned a lot from the game of golf. The father (Steve) now a retired architect, club champion and not long ago a plus handicap golfer and the son (Michael) now an engineering graduate after some years as a golf professional and tour caddie, have collaborated to produce this book. Our collaboration and conversations over the last five years have been extraordinary in many ways. Not many get the opportunity to collaborate in game changing discoveries. Fewer still are father and son. The development of The Science of Putting has been a rewarding experience on many levels. A large number of people have made important contributions to the advancements of putting science now included in this book.

Deane Beman, one of the best putters the game of golf has ever known, has been involved along the entire journey. Steve Jones, US Open Champion, has been very generous with his time and insights. Rod Spittle, PGA Tour Champions player, has spent many hours in the putting lab developing information that has been important to this book. Thanks to all of the Tour players for testing in our lab.

Jeff Ryan, Class A PGA Professional, has devoted endless hours in the lab, in conversations and collaboration that have advanced our understanding of the science of putting. Jeff's passion for the game and understanding of the importance of moment of inertia in putting are in many ways the reason The Science of Putting exists today as a book.

Jim MacKay, One-Putt Enterprises (the Quintic Ball Roll Lab in the USA) was involved in early testing, before our lab was set up, and was involved in the discovery of the weight optimization fitting process. Jim has made himself available for conversations on a regular basis over the last six years. His understanding of launch monitor testing, and the science of putting in general, has been vital to developing many of the subjects of this book.

Hundreds of amateurs, professional golfers, colleagues and friends have played their part in testing, collaborating, discussing, and discovery. Thank you all. Jim McCarthy and Grant Gomes, of Cure Putters, have been friends and collaborators in many ways. Their encouragement, engagement, insights and suggestions have been invaluable.

A note from Steve:
I am the "Science Dude" at CURE Putters and have benefited from innumerable conversations and interactions with friends who have resigned themselves to the fact that a conversation with me is more likely than not to involve the science of putting. Thank you all.

It goes without saying (though almost every author says it) that my wife of 50+ years has been supportive. This has been a consuming passion and my wife Jan has helped in the many ways you would expect, like proofreading innumerable drafts. Most importantly, Jan has been an encouragement, a friend and a blessing. I am thankful.

A note from Michael:
I would not have been able to undertake and complete such a work without the love and support of my family. I have benefited my whole life from Steve and Jan, two of the most loving and supportive parents a child has ever been blessed with. Thank you to Paul and Jennifer, my brother and sister, whose talents and encouragement have always pushed me to be better and to do more.

Thank you to Mike Schnarr who took the time and had to patience to mentor a young man and to be an example of what a true golf professional should be. Thank you to Dr. Tim Frank for his tireless efforts and what can only be described as love in teaching engineering, math, and physics to his students. Finally, to my daughter Brittney, Thank you sweetheart for your unwavering faith and support in my many endeavors. Your beautiful spirit and ceaselessly positive outlook have been an indescribable inspiration for me.

Introduction to the BIG Ideas in the book

The first six sections of the book contain important new discoveries and observations about the modern putting stroke, ball roll physics, putting posture and the modern putter.

1) Putting **TEMPO**

   Tour players exhibit pendulum tempo. They have the same tempo with constant acceleration for ALL putts of ALL lengths. A pendulum putting tempo is the most important putting fundamental, by far; and it is easy to practice, and practice can yield dramatic improvement.

2) **BALL ROLL** physics

   A putted ball is not instantly rolling. It skids, begins to roll and then finally truly rolls on the green. Understanding the relationship of ball speed and putt distance is fundamental.

3) **BACKSTROKE LENGTH PHYSICS** in a modern pendulum putting stroke

   With a well-established personal pendulum tempo (with constant acceleration), backstroke length controls putt distance because it directly controls ball speed.

4) **BACKSTROKE LENGTH ON SLOPING GREENS** to control distance

   The effective speed of greens varies with the slope of the green, and can be estimated quite accurately allowing backstroke length for all putts on sloping greens to be estimated.

5) Putting **POSTURE**

   Most tour players have chosen a two-plane neutral arc putting posture that aligns the axis/plane of spine rotation with the plane of rotation of the putter shaft. Aligning these two axes has two benefits: 1) putter stability from rotation on the plane of the putter shaft; and 2) ease of shoulder rotation in a pendulum putting tempo.

6) Optimize Weight to Maximize **STABILITY (MOI)**

   Maximum MOI reduces twist at impact from off-center hits (increased forgiveness) and MOI increases stability in the stroke itself. Tests have shown that face angle variation at impact can be reduced on the order of 50% by weight optimization with a high MOI putter. The discovery of weight optimization allows every player to putt with their personal maximum putter stability.

The IDEA Icons

IDEA icons have been inserted throughout the book to aid in navigating the main ideas and other important principles and concerns.

for most of the important ideas in the book

for "taking a closer look" at details

for special concerns and warnings

Many ideas in this book book will challenge widely held beliefs; many of the new ideas and discoveries are revolutionary, and provide teachers and coaches of putting many news tools and methods.

NOTE: This book is intended for the use of golf teachers, coaches and inquisitive players of all levels. Its vocabulary is technical and precise. The anecdotes and extended stories typical of popular golf teaching are not included. Brevity and clarity are the goals.

Email Questions and Comments to

DavisPuttingScience@gmail.com

The Science of Putting

Section 1
Pendulum Putting Tempo

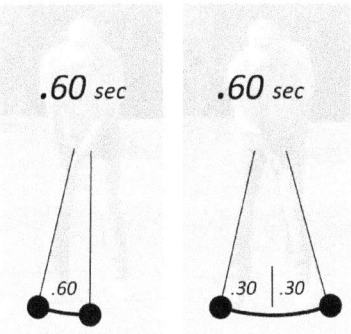

Pendulum Putting Tempo

1.1 The Modern Pendulum Tempo Putting Stroke

The putting stroke of a tour player looks like and behaves like a pendulum. Tour players have the same tempo, cadence and total stroke time for ALL putts of EVERY length. This is a characteristic of pendulums well known to physicists. Natural pendulums behave this way because of the <u>constant acceleration</u> of gravity. Tour players have smooth pendulum like putting strokes because they have smooth constant acceleration throughout their backstroke, across the transition and into forward stroke stroke to impact.

Pendulum Putting Swing Time

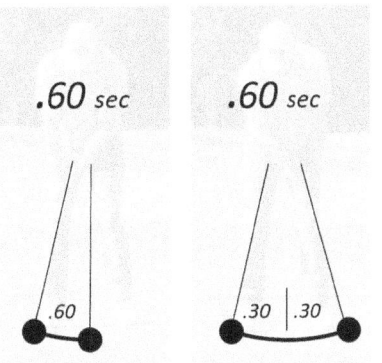

Think of the backstroke as a full swing of a pendulum, taking about 0.60 seconds.

The forward stroke takes the same 0.60 seconds.

Impact is in the middle of the forward stroke, about 0.30 seconds in.

The 2:1 relationship of the backstroke to half the forward stroke is confusing and impossible to feel.

Tour players' backstroke and forward stroke take the same amount of time, about

**0.60 sec
for
ALL putts
of
EVERY length**

~ 0.60 sec backstroke , ~0.60 sec through stroke
<u>Every</u> *Swing the same ~0.60 sec*

Galileo Galilei (1564-1642)

Galileo discovered in 1588 that all pendulums of a given length have the same period (the time for one full cycle). The period of a pendulum does not depend on the weight of the bob, or the amplitude of the swing, only the length of the pendulum matters. Galileo (and all engineers and clockmakers since) would know that all of the various amplitude pendulums shown immediately below have the same period because they have the same length. Weight doesn't matter.

The formula for the period (t) of a pendulum is:

$$t = \pi\sqrt{L/g}$$

The ONLY variable is the length (L) of the pendulum;
π (3.1416) and g (gravity) are constants

The period of a pendulum depends <u>ONLY</u> on the length of the pendulum.
<u>All</u> of the "swings" below whether for a very short putt or a very long putt have the same period.

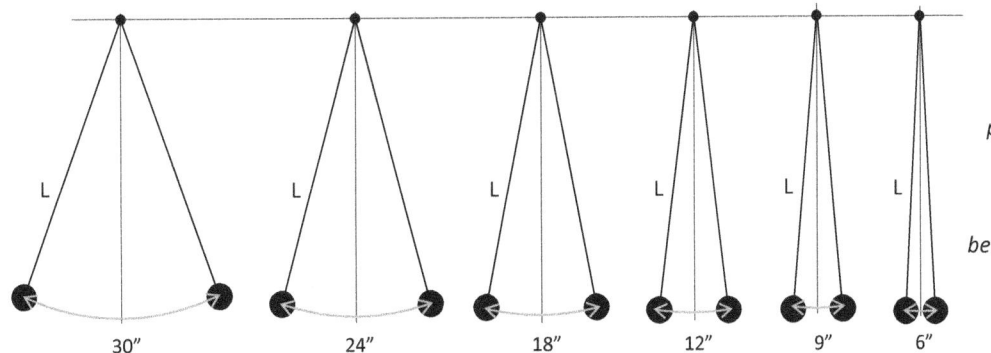

Each of these pendulum swings have exactly the SAME PERIOD (swing time) because they all have the same length

The time is the same for all. The wider strokes above are moving <u>faster</u>. The narrow strokes are moving <u>slower</u>. The force causing this is the <u>constant acceleration of gravity</u>.

The Modern Pendulum Tempo Putting Stroke

The generally accepted rule of thumb is that the backstroke should take about 0.60 seconds and the full forward stroke including the follow through should take about the same 0.60 seconds.

0.00   0.60 sec   1.20 sec

Establishing and practicing a tempo like this is the single most important thing you can do to improve your putting. The myriad faults in amateur putting strokes are almost all related to tempo. Tour players exhibit the constant acceleration of a pendulum in their putting strokes. Amateurs exhibit a wide variety of inconsistent acceleration, deceleration, impact yips, and general inconsistency.

The behavior of a theoretical pendulum can teach us a lot about putting. Notice at the bottom of the previous page that each of these strokes takes the same time, no matter the length of the backstroke and swing. Always the same 0.60 second backstroke and forward stroke; whether for a 3 foot or 60 foot putt. The speed of the strokes increases with the length of the backstroke. This is the nature of a modern pendulum putting stroke with constant acceleration. With a modern pendulum tempo putting stroke with constant acceleration, the timing never varies. Only the backstroke length varies.

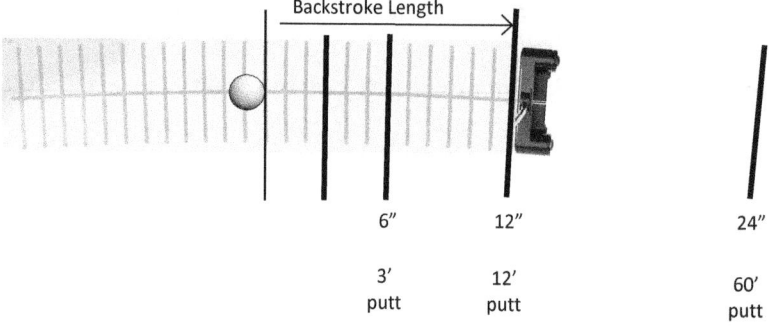

Every additional inch of backstroke produces a longer and longer putt. The physics is quite complicated, but possible to understand. The diagram above shows that a 3' putt will have a backstroke of about 6"; a 12' putt will have a backstroke of about 12"; and a 60' putt will have a backstroke of about 24". This is over simplified, but useful for the general relationship of backstroke length to putt length.

See sections 3 and 4 below for details regarding backstroke length and putt distance.

What follows is an explanation of the physics and principles of the modern pendulum tempo putting stroke; including detailed backstroke length information.

Pendulum Putting Tempo
1.2 The Theoretical Pendulum

A theoretical pendulum has a weightless string; and swings very slowly in comparison to a putting stroke. Shown below, the theoretical pendulum of an average player's height has a 2.40 second period. Remember, a pendulum's period is a full cycle, from beginning point to the other end and then back to the beginning point again. So the 2.40 second period of the theoretical pendulum relates to a backstroke of 1.20 seconds. Stupid slow.

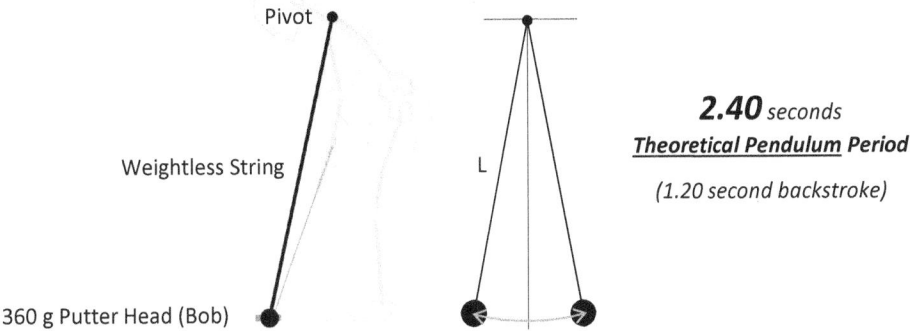

The Second Harmonic
The actual biomechanical putting pendulum does not have a weightless string. The arms and hands weight about 9,798 g (21.6 lbs) and the putter (shaft, grip and putter head) weight about 541 g (1.2 lbs). The arms and hands are so heavy that the effective center of gravity of the biomechanical pendulum assembly is near the elbows. The effective length of the actual pendulum is about ¼ of the length of the theoretical pendulum shown above. A ¼ length theoretical pendulum oscillates at the "second harmonic" of the theoretical pendulum, with a period of 1.20 seconds (0.60 second backstroke).

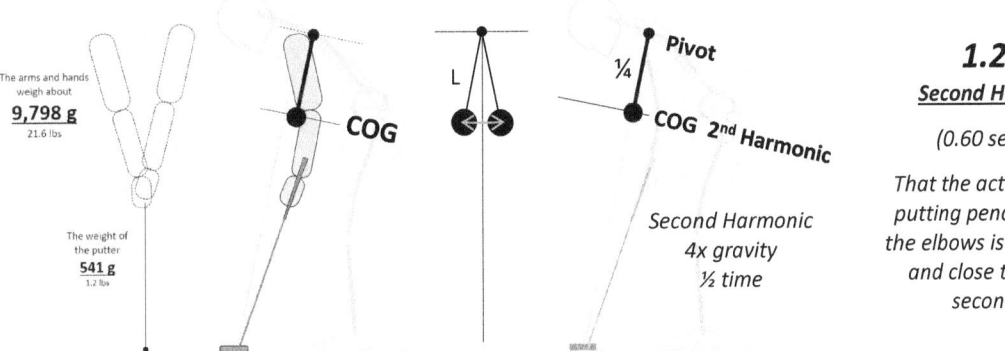

The second harmonic is a natural resonance, and it provides **a more stable (force resistant) pendulum swing**. The second harmonic is very simple to estimate for pendulums of varying lengths.

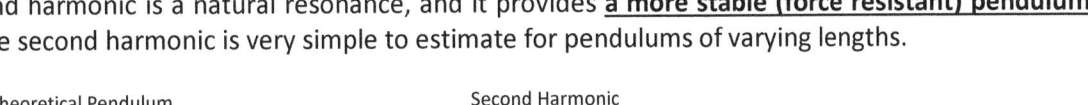

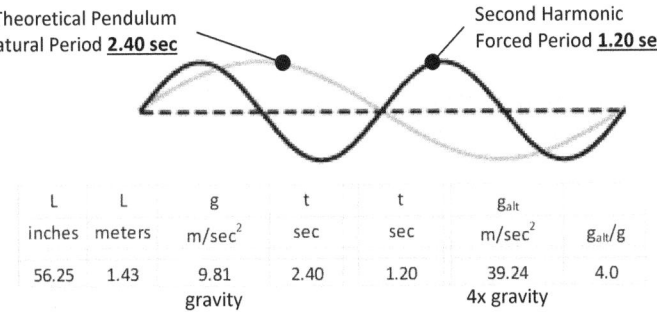

Grober et al have studied and recommended the second harmonic; claiming that this second harmonic is a resonance that the theoretical pendulum "wants" to oscillate at. The second harmonic is a natural resonant frequency that provides a **more stable pendulum swing**.

L inches	L meters	g m/sec^2	t sec	t sec	g$_{alt}$ m/sec^2	g$_{alt}$/g
56.25	1.43	9.81 gravity	2.40	1.20	39.24 4x gravity	4.0

The second harmonic provides a workable estimate of the putting pendulum period.

Second Harmonic Method

This simple method provides a quick reasonably accurate method for determining an approximation of the putting pendulum period for an individual. The pivot of the theoretical putting pendulum is in the spine between the shoulders (near the T3/T4 joint). The period of the second harmonic of the theoretical pendulum is shown in the right column of the table below. The second harmonic period is one half the theoretical pendulum period; the constant acceleration force is FOUR times gravity.

Biometric Second Harmonic Method			4	
Hgt In	Theoretical Pendulum COG inches from pivot inches	Theoretical Pendulum COG meters from pivot meters	Theoretical Pendulum COG Period t sec	Theoretical Pendulum Second Harmonic t/2 sec
53	39.4	1.00	2.01	1.00
54	40.1	1.02	2.02	1.01
55	40.8	1.04	2.04	1.02
56	41.5	1.05	2.06	1.03
57	42.3	1.07	2.08	1.04
58	43.0	1.09	2.10	1.05
59	43.8	1.11	2.11	1.06
60	44.5	1.13	2.13	1.07
61	45.2	1.15	2.15	1.08
62	46.0	1.17	2.17	1.08
63	46.7	1.19	2.19	1.09
64	47.5	1.21	2.20	1.10
65	48.2	1.22	2.22	1.11
66	49.0	1.24	2.24	1.12
67	49.7	1.26	2.25	1.13
68	50.4	1.28	2.27	1.14
69	51.2	1.30	2.29	1.14
70	51.9	1.32	2.30	**1.15**
71	52.7	1.34	2.32	1.16
72	53.4	1.36	2.34	**1.17**
73	54.1	1.38	2.35	1.18
74	54.9	1.39	2.37	1.18
75	55.6	1.41	2.38	1.19
76	56.4	1.43	2.40	**1.20**
77	57.1	1.45	2.42	1.21
78	57.9	1.47	2.43	1.22
79	58.6	1.49	2.45	1.22
80	59.3	1.51	2.46	1.23
81	60.1	1.53	2.48	1.24
82	60.8	1.54	2.49	1.25
83	61.6	1.56	2.51	1.25
84	62.3	1.58	2.52	1.26
85	63.0	1.60	2.54	1.27
86	63.8	1.62	2.55	1.28
87	64.5	1.64	2.57	1.28
88	65.3	1.66	2.58	1.29
89	66.0	1.68	2.60	1.30

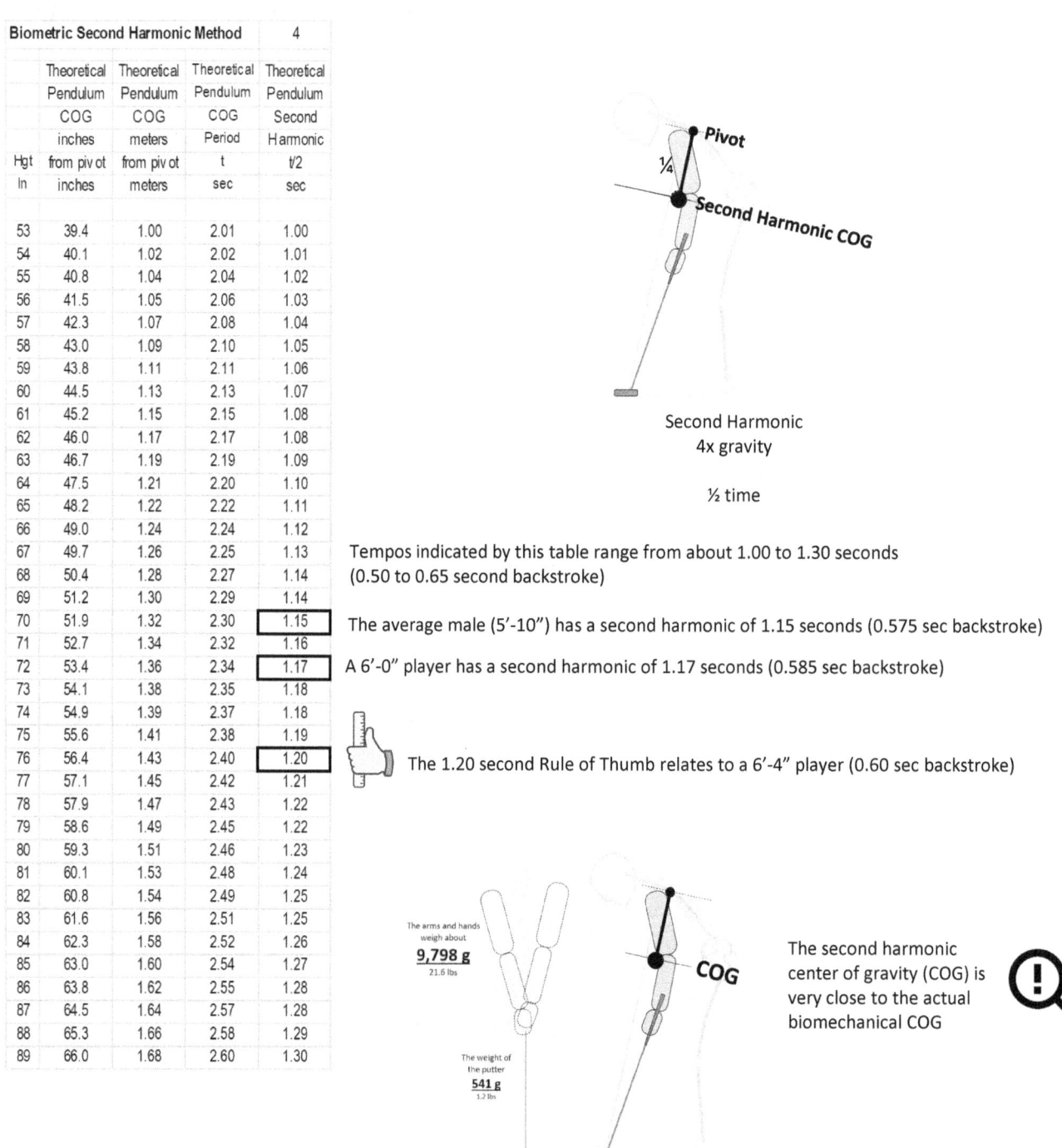

Second Harmonic
4x gravity

½ time

Tempos indicated by this table range from about 1.00 to 1.30 seconds (0.50 to 0.65 second backstroke)

The average male (5'-10") has a second harmonic of 1.15 seconds (0.575 sec backstroke)

A 6'-0" player has a second harmonic of 1.17 seconds (0.585 sec backstroke)

The 1.20 second Rule of Thumb relates to a 6'-4" player (0.60 sec backstroke)

The arms and hands weigh about **9,798 g** 21.6 lbs

The weight of the putter **541 g** 1.2 lbs

The second harmonic center of gravity (COG) is very close to the actual biomechanical COG

The actual "human + putter" pendulum is a very complicated biomechanical system; but not beyond analysis. See the following page.

Body Weight Segment Analysis Method

The relative and absolute weight of the arms and hands dominates the putting pendulum. The details of putter parts weight (and relative weight) are not widely understood. The figures and tables below describe arms and hands weight.

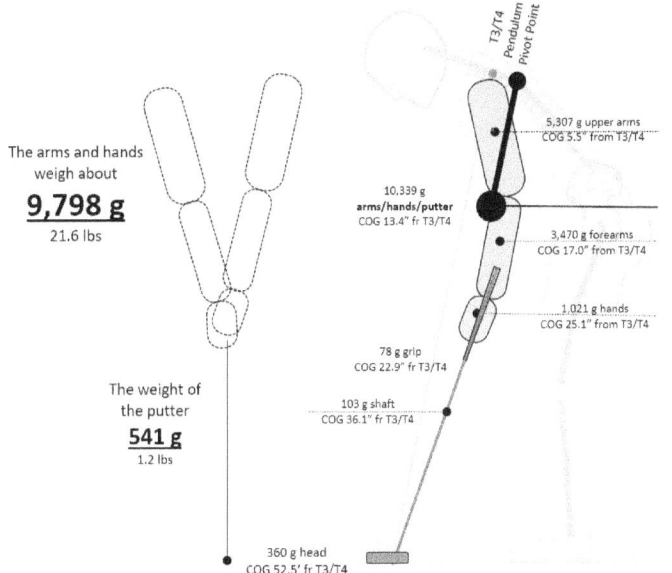

The theoretical pendulum has a weightless string, so the center of gravity is in the bob; and the pivot has no friction so the constant acceleration of gravity causes it to swing smoothly, forever.

The "arms/hands/grip/shaft string" in the actual "human + putter" pendulum is not weightless. Each of the elements has a mass and a center of gravity and the pivot has so much friction that it barely gets to impact from the back of the backstroke (with gravity alone).

NOTE:
The hands alone are almost twice the weight of putter head; each hand about the weight of the head. The arms and hands together dominate the weight of the putter itself

Body Segment as Percent of Total Body Mass

	Harles 1860	Braune & Fisher 1889	Bernstein 1936	Dempster 1955	de Leva 1996	Winter 2009	Range	Average Calc	Value Used
Upper Arms	6.48	6.72	5.26	5.30	5.26	7.20	5.26 - 7.20	5.76	7.20
Forearms	3.62	4.56	3.64	3.10	3.00	4.40	3.00 - 4.56	3.54	4.40
Hands	1.68	1.68	1.26	1.20	1.17	1.40	1.17 - 1.68	1.26	1.40
Total Arms & Hands	11.78	12.96	10.16	9.60	9.43	13.00	9.43 - 13.00	10.55	13.00

NOTE: Average body segment weight percentage used in calculations below are confirmed as reasonable

Putting Pendulum Period based on Center of Gravity of human and putter parts

	Height of Person		72	in		
	Weight of Person		180	lbs	81647	wgt in g
	Putter Length		34	length inches		
	Putter Head Weight		360	wgt grams		
	% total body weight	inches from pivot	g weight	gcm moment	moment = weight x inches from pivot	
upper arms	7.20%	5.5	5,879	32,332	weight = % x total body weight	
forearms	4.40%	17.0	3,592	61,072	weight = % x total body weight	
hands	1.40%	25.1	1,143	28,691	weight = % x total body weight	
Sub Total Arms and Hands			10,614	122,094		
grip		22.9	78	1,786	weight of grip	
shaft		36.1	103	3,718	weight of shaft	
putter head weight		52.4	360	18,864	weight of putter head	
Sub Total Putter Parts			541	24,369		
Total Putting Pendulum	L to COG in inches	13.1	11,155	146,463	0.333	L to COG in meters
	Actual Pendulum t	t	1.16	seconds	$t = 2\pi\sqrt{L/g}$	
Theoretical Pendulum	L to COG in inches	53.40			1.356	meters
	Theoretical Pendulum t	t	2.34	seconds	$t = 2\pi\sqrt{L/g}$	
	Second Harmonic t	t/2	1.17	seconds	$t/2 = \pi\sqrt{L/g}$	

NOTE:
The Putting Pendulum t **(1.16 sec)** is very close to the second harmonic of the Theoretical Pendulum t/2 **(1.17 sec)**

The table on the following page incorporates this data into a Body Weight Segment Analysis Method.

Body Weight Segment Analysis Method

The table below summarizes the distance of the putting pendulum COG by height and weight of the individual. Taller players, and players with light arms, tend to have slower tempo. Heavier players, and players with very heavy arms, tend to have faster tempo.

The biometric second harmonic method is shown in the right column for reference. The 1.20 second rule-of-thumb is clearly not correct for all individuals. The second harmonic method informs the question regarding height of the individual. The body segment method adds weight to the equation; and provides the best estimate of an individual's putting pendulum period.

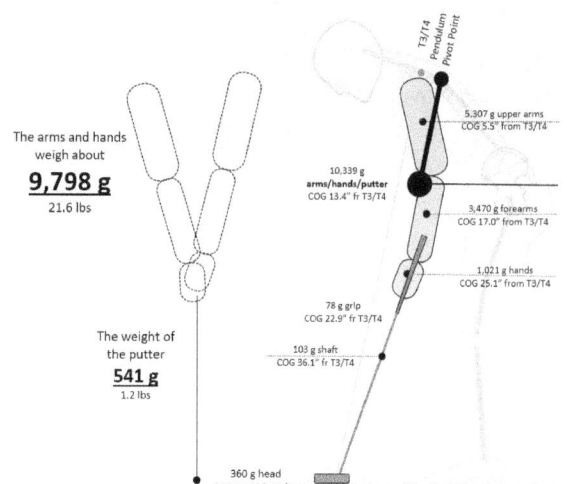

Putting Pendulm Period Analysis by Body Segment Wgt Analysis

hgt in	thin extrema COG inches from pivot	wgt lbs	pendulum period t sec	low average COG inches from pivot	wgt lbs	pendulum period t sec	avg weight COG inches from pivot	wgt lbs	pendulum period t sec	high average COG inches from pivot	wgt lbs	pendulum period t sec	heavy extrema COG inches from pivot	wgt lbs	pendulum period t sec	Biometric Second Harmonic Method t/2 sec
54	11.73	63	1.10	11.42	71	1.08	10.69	100	1.05	10.26	128	1.02	9.87	170	1.00	1.01
55	11.87	65	1.10	11.53	74	1.09	10.81	104	1.05	10.39	133	1.03	10.02	176	1.01	1.02
56	12.01	67	1.11	11.68	76	1.09	10.95	107	1.06	10.53	138	1.04	10.16	182	1.02	1.03
57	12.11	70	1.11	11.80	79	1.10	11.08	111	1.06	10.66	143	1.04	10.30	189	1.03	1.04
58	12.25	72	1.12	11.92	82	1.10	11.21	115	1.07	10.80	148	1.05	10.44	196	1.03	1.05
59	12.36	75	1.12	12.04	85	1.11	11.35	119	1.08	10.94	153	1.06	10.59	202	1.04	1.06
60	12.50	77	1.13	12.17	88	1.12	11.49	123	1.08	11.09	158	1.06	10.74	209	1.05	1.07
61	12.62	80	1.14	12.33	90	1.12	11.62	127	1.09	11.22	164	1.07	10.88	216	1.05	1.08
62	12.73	83	1.14	12.45	93	1.13	11.76	131	1.10	11.37	169	1.08	11.02	224	1.06	1.09
63	12.88	85	1.15	12.59	96	1.13	11.90	136	1.10	11.50	175	1.08	11.17	231	1.07	1.09
64	13.00	88	1.15	12.70	100	1.14	12.03	140	1.11	11.65	180	1.09	11.32	238	1.08	1.10
65	13.12	91	1.16	12.83	103	1.15	12.17	145	1.12	11.79	186	1.10	11.46	246	1.08	1.11
66	13.28	93	1.16	12.97	106	1.15	12.31	149	1.12	11.93	192	1.10	11.60	254	1.09	1.12
67	13.40	96	1.17	13.11	109	1.16	12.45	153	1.13	12.08	197	1.11	11.75	261	1.10	1.13
68	13.53	99	1.18	13.25	112	1.16	12.60	158	1.13	12.22	203	1.12	11.90	269	1.10	1.14
69	13.66	102	1.18	13.37	116	1.17	12.73	163	1.14	12.37	209	1.12	12.05	277	1.11	1.14
70	13.79	105	1.19	13.51	119	1.18	12.87	168	1.15	12.51	216	1.13	12.20	285	1.12	1.15
71	13.93	108	1.19	13.65	122	1.18	13.02	172	1.15	12.65	222	1.14	12.35	293	1.12	1.16
72	14.06	111	1.20	13.78	126	1.19	13.16	177	1.16	12.80	228	1.14	12.49	302	1.13	1.17
73	14.20	114	1.20	13.92	129	1.19	13.30	182	1.17	12.95	234	1.15	12.64	310	1.14	1.18
74	14.34	117	1.21	14.05	133	1.20	13.44	187	1.17	13.09	241	1.16	12.79	319	1.14	1.18
75	14.46	121	1.22	14.19	137	1.20	13.58	193	1.18	13.24	248	1.16	12.94	328	1.15	1.19
76	14.60	124	1.22	14.34	140	1.21	13.73	197	1.18	13.39	254	1.17	13.09	336	1.16	1.20
77	14.74	127	1.23	14.47	144	1.22	13.87	203	1.19	13.53	261	1.18	13.24	345	1.16	1.21
78	14.88	130	1.23	14.61	148	1.22	14.01	208	1.20	13.68	268	1.18	13.39	354	1.17	1.21
79	15.01	134	1.24	14.76	151	1.23	14.16	213	1.20	13.82	275	1.19	13.54	363	1.18	1.22
80	15.15	137	1.24	14.89	155	1.23	14.30	219	1.21	13.97	282	1.20	13.69	373	1.18	1.23
81	15.30	140	1.25	15.03	159	1.24	14.45	224	1.22	14.12	289	1.20	13.84	382	1.19	1.24
82	15.43	144	1.26	15.17	163	1.25	14.60	230	1.22	14.27	296	1.21	13.99	392	1.20	1.25
83	15.58	147	1.26	15.31	167	1.25	14.74	235	1.23	14.42	303	1.21	14.14	401	1.20	1.25
84	15.71	151	1.27	15.46	171	1.26	14.89	241	1.23	14.57	311	1.22	14.29	411	1.21	1.26
85	15.84	155	1.27	15.60	175	1.26	15.03	247	1.24	14.72	318	1.23	14.45	421	1.22	1.27
86	15.99	158	1.28	15.74	179	1.27	15.18	253	1.25	14.87	326	1.23	14.60	431	1.22	1.27
87	16.13	162	1.28	15.88	184	1.27	15.33	259	1.25	15.02	333	1.24	14.75	441	1.23	1.28
88	16.27	166	1.29	16.02	188	1.28	15.47	265	1.26	15.17	341	1.25	14.90	451	1.23	1.29

Rule of Thumb 1.20 seconds

Note: The 1.20 sec Rule of Thumb is too slow for most players

The Tempo indicated for an average weight 6'-0" player is **1.17** seconds (Second Harmonic) and **1.16** seconds (Body Segment Weight Analysis Method).
The **1.20** seconds tempo is indicated only at the Thin Extrema of body weight (at 6'-0" height).

A Body Weight Segment Method Footnote

A heavier body (with heavier arms and hands) will move the theoretical pendulum COG higher. A heavier putter will move the theoretical pendulum COG lower.

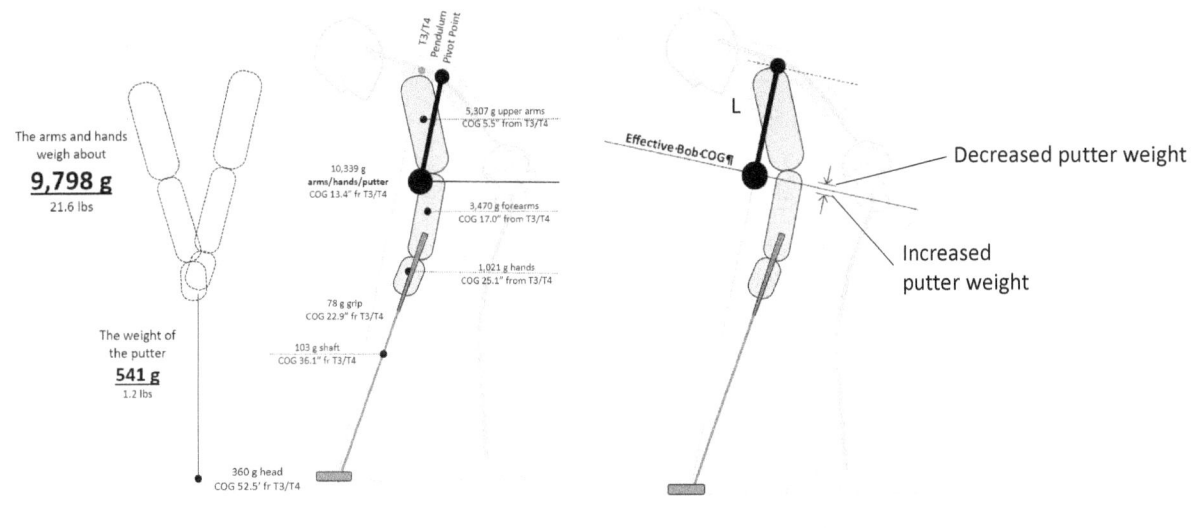

Putting Pendulum Period based on Center of Gravity of human and putter parts

	Height of Person		72	in		
	Weight of Person		180	lbs	81647	wgt in g
	Putter Length		34	length inches		
	Putter Head Weight		360	wgt grams		
	% total body weight	inches from pivot	g weight	gcm moment	moment = weight x inches from pivot	
shoulders/upper thorax	20.00%	4.5	16,329	73,482	weight = % x total body weight	
upper arms	7.20%	5.5	5,879	32,332	weight = % x total body weight	
forearms	4.40%	17.0	3,592	61,072	weight = % x total body weight	
hands	1.40%	25.1	1,143	28,691	weight = % x total body weight	
Sub Total Arms and Hands and shoulders			26,943	195,576		
grip		22.9	78	1,786	weight of grip	
shaft		36.1	103	3,718	weight of shaft	
putter head weight		52.4	360	18,864	weight of putter head	
Sub Total Putter Parts			541	24,369		
Total Putting Pendulum	L to COG in inches	8.0	27,484	219,945	0.203	L to COG in meters
	Actual Pendulum t	t	0.90	seconds		$t = 2\pi\sqrt{L/g}$
Theoretical Pendulum	L to COG in inches	53.40			1.356	meters
	Theoretical Pendulum t	t	2.34	seconds		$t = 2\pi\sqrt{L/g}$
	Second Harmonic t	t/2	1.17	seconds		$t/2 = \pi\sqrt{L/g}$

Including the rotating thorax and shoulders in the body weight segment method calculation (table here) decreases the pendulum period to about 0.90 seconds (0.45 second backstroke). About 2 SD faster than the tour average (see table on the next page). The rotating thorax and shoulders supply important inertia in the putting stroke (increasing impact ratio importantly).

Even adding the 70 gram SAM PuttLAB sensor device to the putter shaft will move the theoretical pendulum COG lower. Increasing the putter head weight does the same thing. Both lengthen the effective pendulum and increases the period of the pendulum by a small amount.

Pendulum Putting Tempo
1.3 The Range of Pendulum Putting Tempos

Putting Stroke Tempo of PGA Tour Players

In 2007, Christian Marquardt, Science&Motion Sports GmbH (SAM PuttLab) collected reference data for 99 PGA Tour Players and published the data in "The SAM PuttLab: Concept and PGA Tour Data". The tempo data is summarized in the table below.

SAM PuttLab PGA Tour Data Analysis
Data of 99 PGA Tour players at 9 PGA tournaments between 2003 and 2005
For a straight and level 4m (13.1 ft) putt (on slower greens the distance was slightly reduced)
Reference target putter speed at impact was 1.5m/s (3.36 mph)

	normal distribution								
	-3SD	-2SD	-1SD	Group Average	+1SD	+2SD	+3SD	Group SD	Individual SD
Backswing Time (BST) seconds	0.40	0.49	0.58	0.67	0.76	0.85	0.94	0.09	0.03
			Tiger Woods	0.660			0.977	Loren Roberts	
Time to Impact (TI) seconds	0.21	0.25	0.28	0.32	0.35	0.39	0.42	0.04	0.01
BST/TI tempo	1.89	1.98	2.06	2.11	2.16	2.20	2.23		
Backswing Metronome BPM	150	122	103	90	79	71	64		

Note the following:
1) The Backswing Time average and SD range of; 0.67 sec average
2) The Time to Impact average and SD range; average 0.32 sec
3) The BST/TI tempo average and SD range of; average 2.11:1
4) The Backswing Metronome BPM average and SD range; average 90 bpm

The PGA Tour average backstroke time, at 0.67 sec, is a little slower than the 0.60 second rule-of-thumb and the calculated theoretical pendulum times indicated in the analysis above. The fact that PGA Tour players average 0.67 seconds could be an indication that the theoretical pendulum calculations are arguing for too fast tempos.

It is also possible that some (perhaps many) players could improve with the faster theoretical pendulum tempos.

Average Tempo of PGA Tour Players in context

The Body Weight Segment and Second Harmonic methods both argue for backstroke times slightly less than the generally accepted rule of thumb 0.60 seconds for most players. A range of observed and suggested tempos is shown in the table below.

		Full stroke seconds	Backstroke seconds	Metronome BPM	
1)	Body Weight Segment Analysis method	1.16	0.580	103	
2)	The Second Harmonic method	1.17	0.585	102	
3)	**Rule of Thumb**	**1.20**	**0.600**	**100**	Rule of Thumb
4)	PGA Tour Average (per SAM PuttLAB)	1.34	0.670	90	
5)	LPGA Tour Average	1.60	0.800	75	
6)	Amateur golfers	1.80	0.900	67	

Examples 1 & 2 above are for a 6' 0" 177 lb player.

The Range of PGA Tour Players Tempo in context

The range from the previous table is in the left column of the detailed table below:

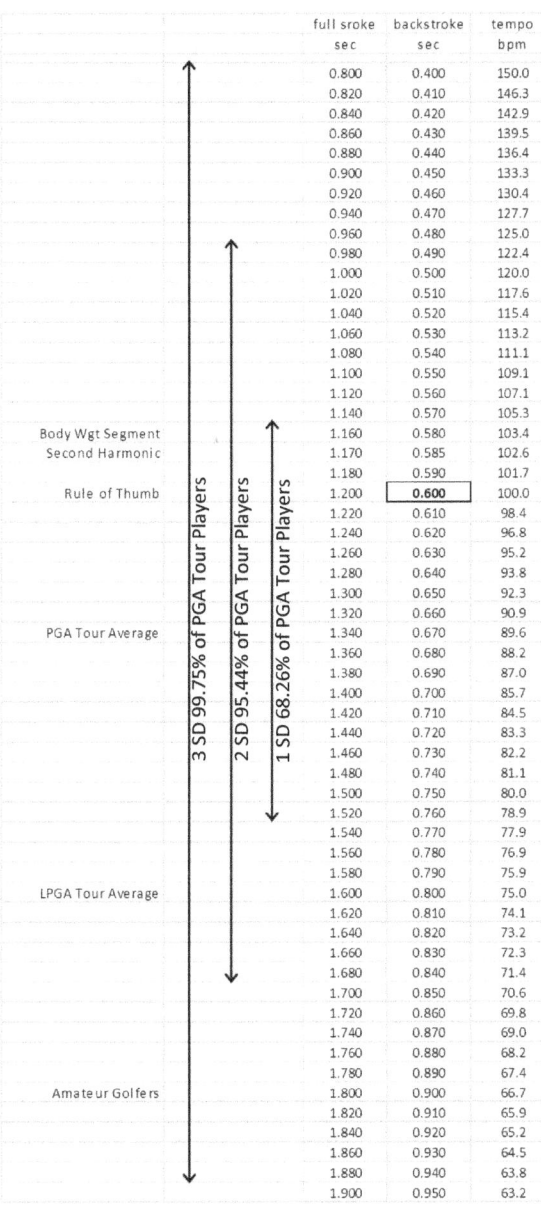

The range of PGA Tour tempos
The average of the PGA Tour is a single point in the range in the table above; but within this data point there is a wide range. See the 1, 2 and 3 standard deviation (SD) ranges shown with the vertical arrows in the table to the left.

There would be a similar range for each of the points above; the PGA Tour range is known thanks to the SAM PuttLAB data.

The big point here is that even the Body Weight Segment method and Second Harmonic method average data point is within the one standard deviation range of the PGA Tour; and therefore not an outlier.

Personal Pendulum Putting Tempo Strategy
Look at the tempo indicated for your personal height and weight by the Second Harmonic and the Body Weight Segment Methods. If they feel too fast, look at the rule of thumb or the PGA Tour average tempo. You should experiment with backstroke times from 0.550 to 0.700 seconds.

Taller players will be slower. Shorter players will be faster. Lighter players will be slower. Heavier players will be faster. You can confirm your personal tempo on a putting launch monitor or using a metronome.

The first priority should be confirming a tempo generally in line with the rule of thumb. Most amateurs (and many tour players) have tempos much slower than 1.20 seconds. Many amateurs are much much too slow with tempos in excess of 1.50 or even 2.00 seconds. The result is less than constant acceleration; often resulting in either deceleration or a yippy catch up acceleration jerk at impact. Amateur players' backstroke time varies from about 0.40 seconds to well over 1.00 seconds. The average backstroke time is likely about 0.900 seconds. The individual variation for amateurs often exceeds 0.20 seconds; 10 to 20 times the variation of tour players.

The focus on the 2:1 putting launch monitor metric has obscured the simple pendulum rhythm of the putting stroke. Most people are surprised to learn that all strokes for all putts of all lengths take the same time – about 1.20 seconds.

PGA Tour players backstroke time averages about 0.67 seconds and varies from about 0.50 seconds to 0.85 seconds (2 SD); for <u>ALL</u> putts of <u>EVERY</u> length. Individual variation in stroke time of a tour player is about 0.01 to 0.03 seconds. VERY consistent.

Pendulum Putting Tempo
1.4 Constant Acceleration is Key

Acceleration is the KEY to a smooth Pendulum Putting Stroke

The natural point of maximum speed being 0.335 (or 0.670/2) seconds into the forward stroke. The Tour players hit the ball 0.32 seconds into the forward stroke, with acceleration.
Even with acceleration continuing after impact.

Tour Tempo Constant Acceleration Visualized

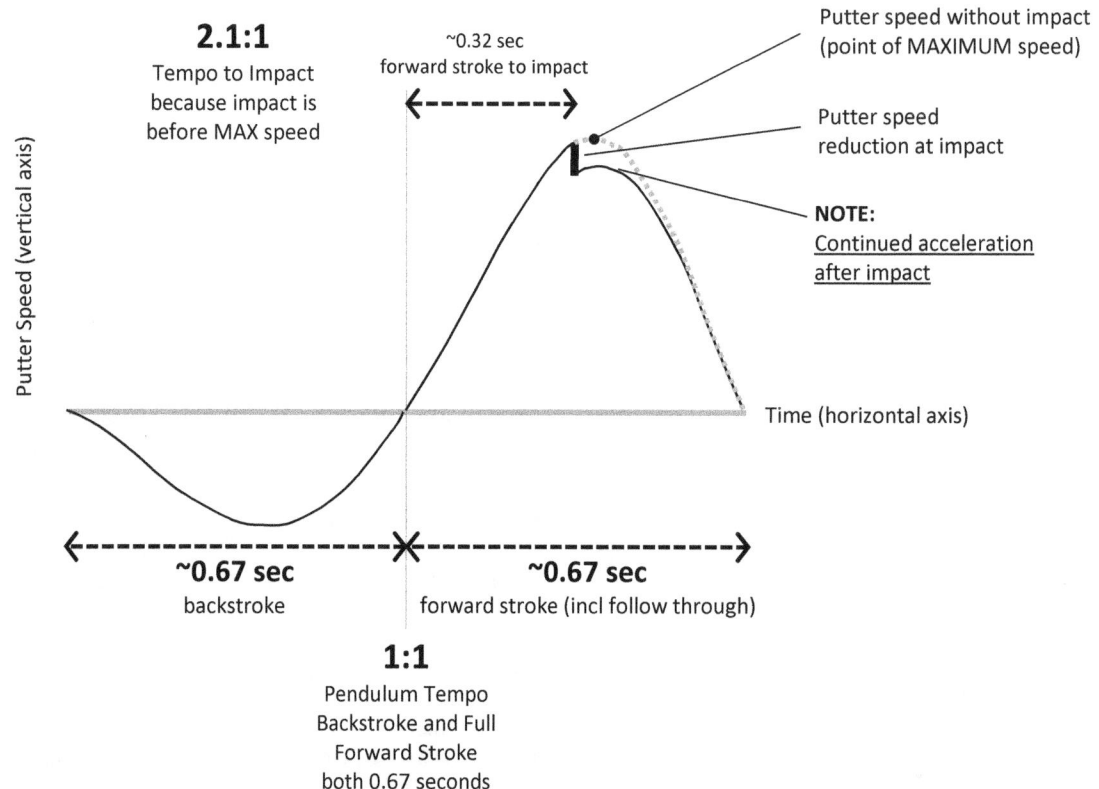

Constant acceleration is <u>CONSTANT</u> acceleration
Note that constant acceleration is not the same as constant velocity. Amateurs often try for constant velocity at impact, focusing on putter speed. Focus on speed often results in deceleration, or jerky acceleration. There is NEVER constant velocity in a tour player's putting stroke. There is only constant acceleration. Acceleration is the key to a pendulum putting tempo. Constant acceleration, like gravity multiplied.

An example of intuitive constant acceleration
The underhand toss is a well-known integrated motion involving intuitive constant acceleration from the beginning of the backswing through the toss. The putting pitfall of slow backswing and jerky acceleration panic at the toss is intuitively avoided in the understand toss. Smooth constant acceleration to the toss.

From child's play to the athletic toss of a professional baseball player, an **underhand toss** will have a flowing wind up, transition, weight shift and release with constant acceleration and smoothly increasing force.

The tour players' secret move
A natural pendulum accelerates across the transitions at each end of the swing. It is not obvious, but acceleration continues across the transition from backstroke to forward stroke. Think of the apparent deceleration in the backstroke as being caused by acceleration in the forward stroke direction rather than applying the backstroke brakes. In effect, most tour players begin the forward stroke in backstroke portion of the transition and then flow into acceleration in the forward stroke. This is the secret of a smooth accelerating putting stroke.

The "feeling" is a little aggressive, a little abrupt at the back of the backstroke; not allowing the putter to come to anything like rest. As the putt length grows from a short putt to longer and longer putts, the force required to achieve constant acceleration, and the pendulum tempo, grows quite large. The "feeling" for long putts is quite aggressive.

The putting pendulum tempo (period), like the natural pendulum tempo (period), does not vary. ALL swings of ALL lengths for ALL putts of ALL lengths are completed in the same time period, about 1.20 seconds (0.60 sec backstroke).

The 2:1 Putting Tempo Idea has been a confusing distraction (and impossible to feel)
The focus on the fact that the first half of the forward swing is about 0.30 seconds long in relation to the full backstroke being 0.60 seconds backstroke, has obscured the fact that both the full backstroke and the full forward swing (including the follow through) are both about 0.60 seconds long.

The timing of the full backstroke equal to the timing of the full forward stroke is easy to feel, and practice. The 2:1 launch monitor metric will likely survive because it is difficult to measure the end of the follow through with a launch monitor (no simple point of rest like address).

The Most Common Amateur Error
The most common amateur error combines a too long backstroke length with poor tempo; most often with deceleration at impact. The putting strokes and tempos of amateur players vary a lot because their acceleration varies so much. The key to the smooth swing of a natural pendulum and the pendulum putting tempo of tour players is constant acceleration. Not just acceleration at impact, but **constant acceleration** in the entire stroke. Tour players achieve their precision because of constant acceleration. Not just acceleration at impact, but **constant acceleration** in the entire stroke. Improved distance control (and better contact) is possible by combining appropriate backstroke length with good putting tempo. Backstroke length must vary with putt length. Putting tempo must remain constant.

The Most Important Putting Fundamental
A smooth, constantly accelerating, pendulum tempo is by far the most important putting fundamental. Nothing else even comes close. Work on your pendulum tempo like your life depended on it. The quality of your life as a golfer does depend on it.

Pendulum Putting Tempo
1.5 Practicing the Putting Pendulum Tempo

Pendulum Putting Tempo Metronome Practice
The metric rhythm, cadence, of the pendulum putting stroke can be established and reinforced by using a metronome. Download a metronome app and set it for the beats per minute (bpm) prescribed for your personal pendulum putting tempo in the table below. A 0.60 second tempo is 100 bpm.

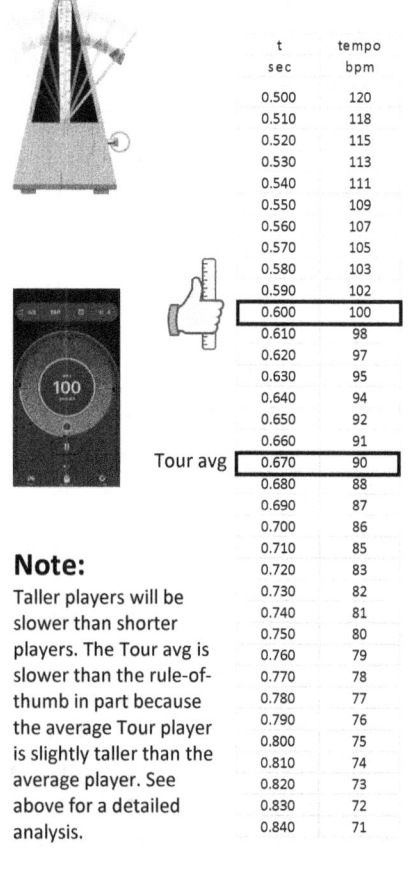

	t sec	tempo bpm
	0.500	120
	0.510	118
	0.520	115
	0.530	113
	0.540	111
	0.550	109
	0.560	107
	0.570	105
	0.580	103
	0.590	102
	0.600	100
	0.610	98
	0.620	97
	0.630	95
	0.640	94
	0.650	92
	0.660	91
Tour avg	0.670	90
	0.680	88
	0.690	87
	0.700	86
	0.710	85
	0.720	83
	0.730	82
	0.740	81
	0.750	80
	0.760	79
	0.770	78
	0.780	77
	0.790	76
	0.800	75
	0.810	74
	0.820	73
	0.830	72
	0.840	71

Note:
Taller players will be slower than shorter players. The Tour avg is slower than the rule-of-thumb in part because the average Tour player is slightly taller than the average player. See above for a detailed analysis.

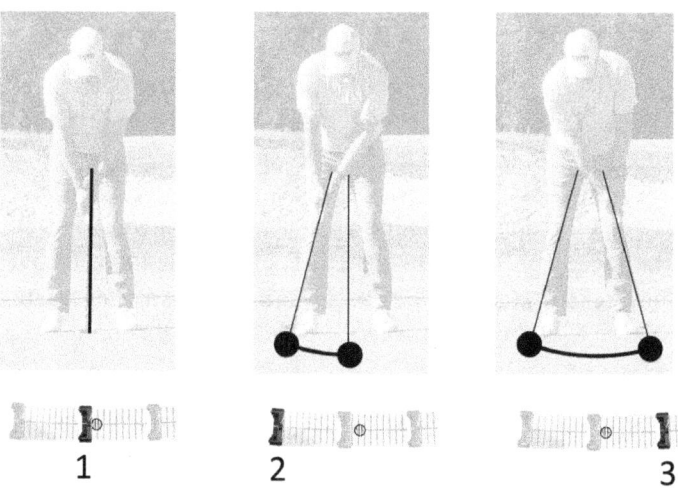

1 **2** **3**

All of the figures here show a right-handed player. The figure above can be used by left-handed players as if looking in a mirror. The figure below may be helpful for a right-handed player to visualize the putter path and tempo at his feet

Calibrate your putting stroke to the metronome.

Count 1,2,3.

Start your backstroke on (1).
End Backstroke and start your forward stroke on (2)
Finish your follow through on (3)

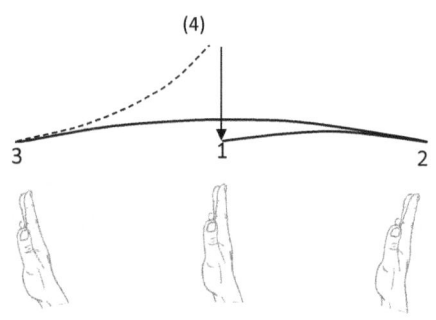

You can "feel" the tempo in the hand/baton movements (left) of an orchestra conductor.

FEEL the downstroke (1) as the "cue" to start the stroke moving smoothly to 2 and 3 on the tempo.

Use (4) to repeat.

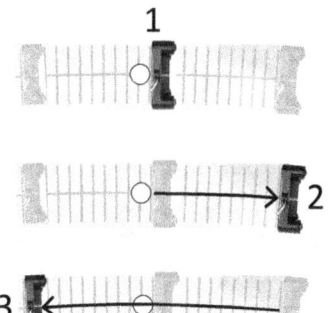

Reaction time for sounds averages about 0.15 seconds; but is dramatically reduced for regular metric rhythms like a metronome.

Putting with a Pendulum Putting Tempo is liberating and easy. You will find yourself making a more confident stroke with better impact and consistency.

You can WALK to your personal pendulum putting tempo

If you are reasonably fit, you can almost certainly "walk" to your Personal Putting Pendulum Tempo; with a footfall on each beat. This will be a relatively fast walk. With a metronome, you can walk with right, left, right, left falling on the beats. Once you know how much "hurry" you need in your step to hit your tempo, you can use it to calibrate your tempo, anytime; during a round or during practice. The "hurry" in your step for this drill is similar to the hurry required in your putting stroke to complete the full stroke in the tempo.

REMEMBER!
Every stroke ALWAYS takes the same time, about 0.60 seconds
No matter how long; no matter how short; ALWAYS the same.
EVERY backstroke of ANY length takes 0.60 seconds; EVERY through stroke takes the same 0.60 seconds.

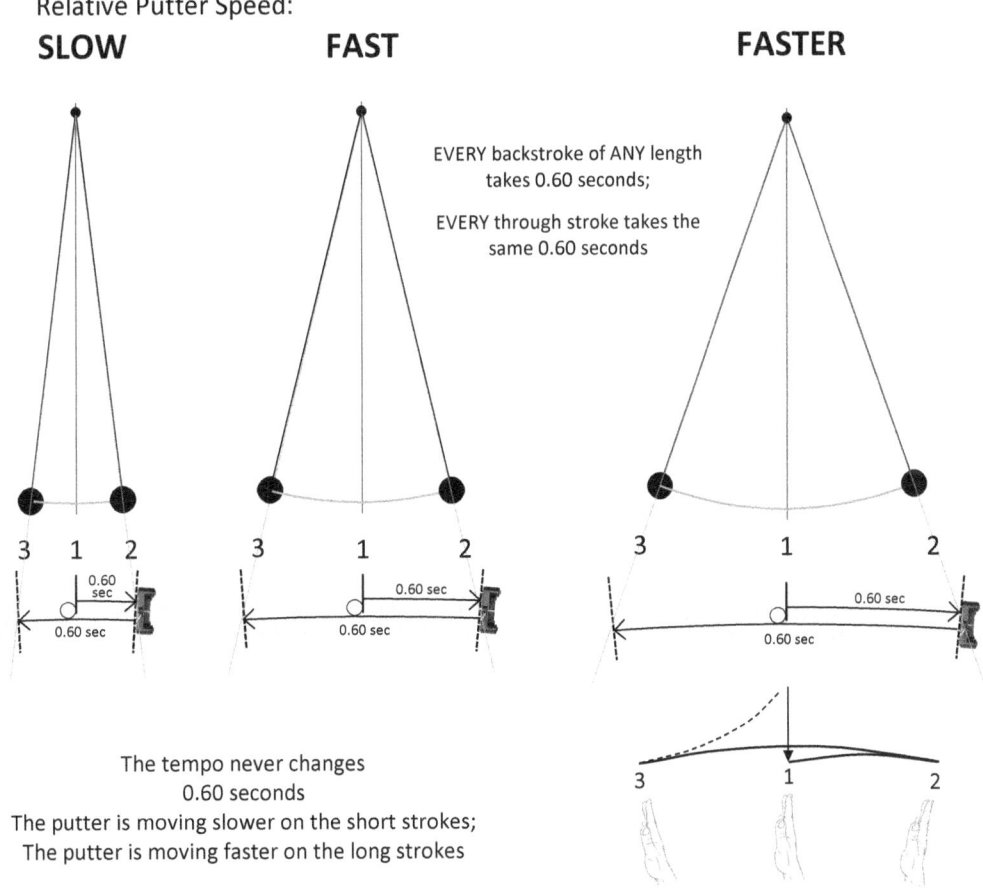

Some will find the longer strokes harder to make on tempo;
most find the very short strokes harder to make on tempo.

With practice you can learn to develop a pendulum putting stroke and tempo with constant acceleration for all strokes of all lengths.

Chip shots of less than 25 yards have the same pendulum tempo (0.60 second backstroke and 0.60 second through stroke). You should explore your personal putting and chipping tempos together. They will reinforce each other.

Practice is as simple as 1,2,3

Section 2
Ball Roll Physics

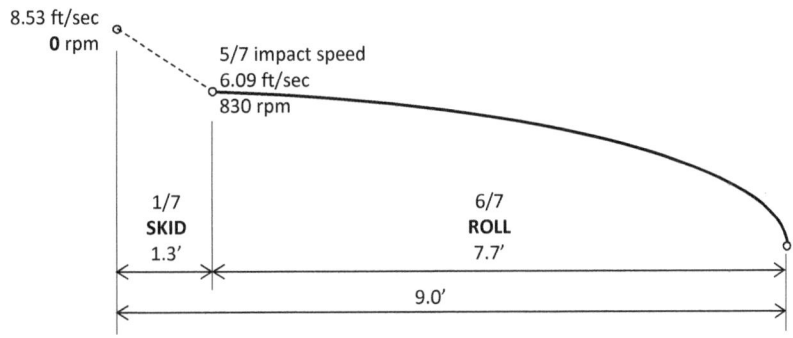

Ball Roll Physics
2.1 Launch, Skid, Roll and Putt Distance

The previous sections have looked at the physics of the putter pendulum. This section will look at the ball and how it rolls on the green. The next section will rejoin the putting stroke pendulum analysis with an analysis of backstroke length required to get the ball started at the required speed to produce the desired total putt distance. But first we need to understand how the ball behaves on the green.

The ball at impact is a rest, zero rpm. The transition from no roll to pure roll is the skid phase of the putt. A ball is skidding if it has less forward roll rpm than pure roll. The launch conditions vary, but the distance a putt finally rolls on a green is less impacted by launch/spin than generally understood. Even very significant topspin decreases the skid distance only very little; and very significant backspin increases the skid distance, very little. Similarly, an initial flight of a few inches doesn't change the total flight/skid distance very much. It is reasonable therefore to assume an optimal launch.

The skid roll constants:
The skid distance will be about 1/7 of the total putt length.
The roll phase will be about 6/7 of the total putt length.
The ball speed after skid, at the beginning of the roll phase, will be about 5/7 of the initial ball speed.

The putted ball starts with ZERO forward roll rpm. The energy consumed in transforming skid (zero rpm) with no roll to true roll is quite large. Putt A (immediately below) starts at 8.53 ft/sec and skid/rolls a total of 9'.

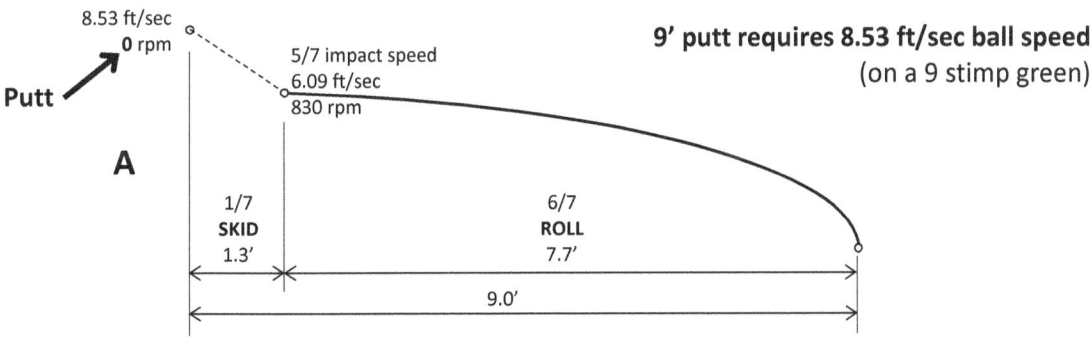

A putt with an initial ball speed of 8.53 ft/sec (5.82 mph) will produce a putt that rolls a total of 9' on a 9 stimp green.

The arcane **ball speed of 8.53 ft/sec (5.82 mph) produces the stimp length putt** on greens of any stimp:

8.53 ft/sec (5.82 mph) on	6 stimp =	6' putt
8.53 ft/sec (5.82 mph) on	7 stimp =	7' putt
8.53 ft/sec (5.82 mph) on	8 stimp =	8' putt
8.53 ft/sec (5.82 mph) on	9 stimp =	9' putt
8.53 ft/sec (5.82 mph) on	10 stimp =	10' putt
8.53 ft/sec (5.82 mph) on	11 stimp =	11' putt
8.53 ft/sec (5.82 mph) on	12 stimp =	12' putt
8.53 ft/sec (5.82 mph) on	13 stimp =	13' putt
8.53 ft/sec (5.82 mph) on	14 stimp =	14' putt

The concepts of launch, skid and roll are crucial to understanding the behavior of a putted ball on the green. But a putted ball is not like a ball rolling off the bottom of a stimpmeter. See the next page for a detailed explanation.

Ball Roll Physics
2.2 Stimpmeter Physics – Bounce and Roll

A putted ball behaves very differently from a ball rolling off a stimpmeter.

The stimpmeter is inclined at angle of 20 degrees and slow-motion video analysis of the interaction of the ball and the ground shows the ball bouncing several times before it begins to roll. Five bounces on a 9 stimp green were clearly evident and very little speed was consumed in the bouncing phase because the ball contacted the ground with significant top spin. <u>The ball starting at a speed of 6.00 ft/sec and 818 rpm at the bottom of a stimpmeter rolls a total of 9' on a 9' stimp green.</u>

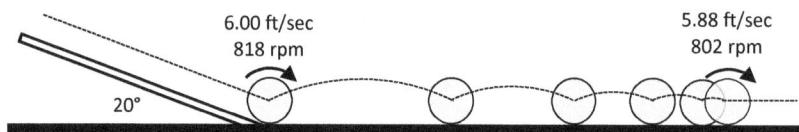

The ball bounces on the ground (several times) due to the the 20° stimpmeter angle.

The ball bouncing off the bottom of a stimpmeter maintains nearly all of its initial velocity through the entire bouncing phase.

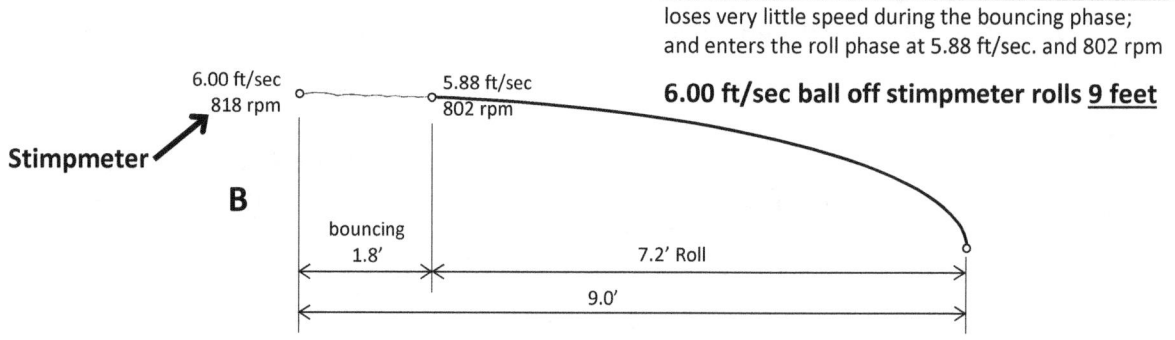

The stimpmeter ball starts with SIGNIFICANT top spin; loses very little speed during the bouncing phase; and enters the roll phase at 5.88 ft/sec. and 802 rpm

6.00 ft/sec ball off stimpmeter rolls <u>9 feet</u>

Putt C (below) starts at the same speed as the stimpmeter ball (but with zero rpm) and skid/rolls only 4.4'. Some very smart people have missed this and assumed a 6 ft/sec putt goes 9'.

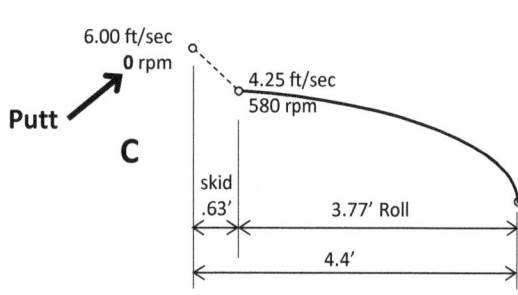

NOTE: The difference between the stimpmeter ball and the putted ball both starting at the same 6.00 ft/sec ball speed.

The stimpmeter ball has 818 rpm of topspin. The putted ball has no spin.

6.00 ft/sec putt only skid/rolls <u>4.4 feet</u>

Note the ball roll distances at various speeds on the charts on the following pages. The energy imparted to the ball varies with the square of the putter speed, creating the parabolic curves. Note the stimpmeter ball ball and 8.53 ft/sec (5.82 mph) ball speed overlaid on the charts.

Ball Roll Physics
2.3 Ball Speed and Stimp determine Putt Distance

How far a putt will roll (on a level green) is mainly determined by two variables: ball speed and green speed (stimp). You control ball speed by controlling the speed of the putter at impact.

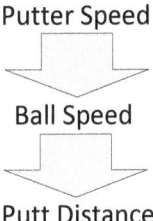

Putter Speed
Ball Speed
Putt Distance

The putter Impact Ratio (Smash Factor with other clubs) varies from about 1.65 to 1.85. Impact Ratio in robot testing is lower than when humans putt. Important variables include face material (steel, aluminum vs soft inserts), shaft stiffness and grip pressure. Softer face inserts produce lowers impact ratios. Stiffer shafts produce higher impact ratio.

The average for humans putting is about 1.75, but can vary a lot.

The skid roll constants:
The skid distance will be about 1/7 of the total putt length.
The roll phase will be about 6/7 of the total putt length.
The ball speed after skid, at the beginning of the roll phase, will be about 5/7 of the initial ball speed.

A complex algorithm can calculate a combination of many putt variables: impact ball flight (launch angle and spin); skid and finally roll. The formula for roll distance, after the initial skid/launch of about 1/7 of the total putt distance, only has two variables: ball speed (v) and coefficient of rolling friction (μ). Gravity (g) is constant. The ball speed (v) is squared. The formula for roll distance (x) is:

$$x = \frac{v^2}{2\mu g}$$

The coefficient of rolling friction (μ) varies with the stimp of the green

$$\text{Coefficient of rolling friction} \quad \mu_r = \frac{7}{5}\left(\frac{0.48}{stimp}\right)$$

stimp ft	μ coefficent
6	0.112
7	0.096
8	0.084
9	0.075
10	0.067
11	0.061
12	0.056
13	0.052
14	0.048

A simplified formula for total putt distance, including skid/launch is:

$$total\ putt\ distance = \frac{v^2}{2\mu g}\frac{7}{6}$$

The data presented in the following tables and graphs, based on these algorithms, has been experimentally verified with hundreds of putts on putting launch monitors and slow-motion video analysis on level greens with a variety of stimps.

Table of Roll Distance (mph ball speed)

stimps	6	7	8	9	10	11	12	13	14	15	16
0.00	0.00	0.00	0.00	0.00	0.00	0.00	0.00	0.00	0.00	0.00	0.00
1.00	0.18	0.21	0.24	0.27	0.30	0.33	0.36	0.38	0.41	0.44	0.47
2.00	0.71	0.83	0.95	1.07	1.18	1.30	1.42	1.54	1.66	1.78	1.89
3.00	1.60	1.86	2.13	2.40	2.66	2.93	3.20	3.46	3.73	4.00	4.26
4.00	2.84	3.32	3.79	4.26	4.74	5.21	5.68	6.16	6.63	7.10	7.58
5.00	4.44	5.18	5.92	6.66	7.40	8.14	8.88	9.62	10.36	11.10	11.84
6.00	6.39	7.46	8.52	9.59	10.66	11.72	12.79	13.85	14.92	15.98	17.05
7.00	8.70	10.15	11.60	13.05	14.50	15.95	17.41	18.86	20.31	21.76	23.21
8.00	11.37	13.26	15.16	17.05	18.94	20.84	22.73	24.63	26.52	28.42	30.31
9.00	14.39	16.78	19.18	21.58	23.98	26.37	28.77	31.17	33.57	35.96	38.36
10.00	17.76	20.72	23.68	26.64	29.60	32.56	35.52	38.48	41.44	44.40	47.36
11.00	21.49	25.07	28.65	32.23	35.82	39.40	42.98	46.56	50.14	53.72	57.31
12.00	25.57	29.84	34.10	38.36	42.62	46.89	51.15	55.41	59.67	63.94	68.20
13.00	30.01	35.02	40.02	45.02	50.02	55.03	60.03	65.03	70.03	75.04	80.04
14.00	34.81	40.61	46.41	52.22	58.02	63.82	69.62	75.42	81.22	87.03	92.83
15.00	39.96	46.62	53.28	59.94	66.60	73.26	79.92	86.58	93.24	99.90	106.56
16.00	45.47	53.04	60.62	68.20	75.78	83.35	90.93	98.51	106.09	113.67	121.24
17.00	51.33	59.88	68.44	76.99	85.55	94.10	102.65	111.21	119.76	128.32	136.87
18.00	57.54	67.13	76.72	86.31	95.91	105.50	115.09	124.68	134.27	143.86	153.45
19.00	64.11	74.80	85.49	96.17	106.86	117.54	128.23	138.91	149.60	160.29	170.97
20.00	71.04	82.88	94.72	106.56	118.40	130.24	142.08	153.92	165.76	177.60	189.44

Ball Speed at Impact (mph)

Only 6 - 16 stimps are displayed here in the tables and graphs. The algorithm allows the calculation of putt distance from ball speed on any stimp green.

There are several things to note on the following graphs. First note the length of putts for a ball speed equal to the ball speed at the bottom of a stimpmeter. This is important. Some very smart people have assumed that a ball rolling of the bottom of a stimpmeter and a putted ball behave the same on the green. They do not. Note the speed of a putted ball (8.53 ft/sec or 5.82 mph) that rolls the distance of the stimpmeter rating. Note the parabolic curve caused by increasing velocity (squared).

Ball Speed in mph

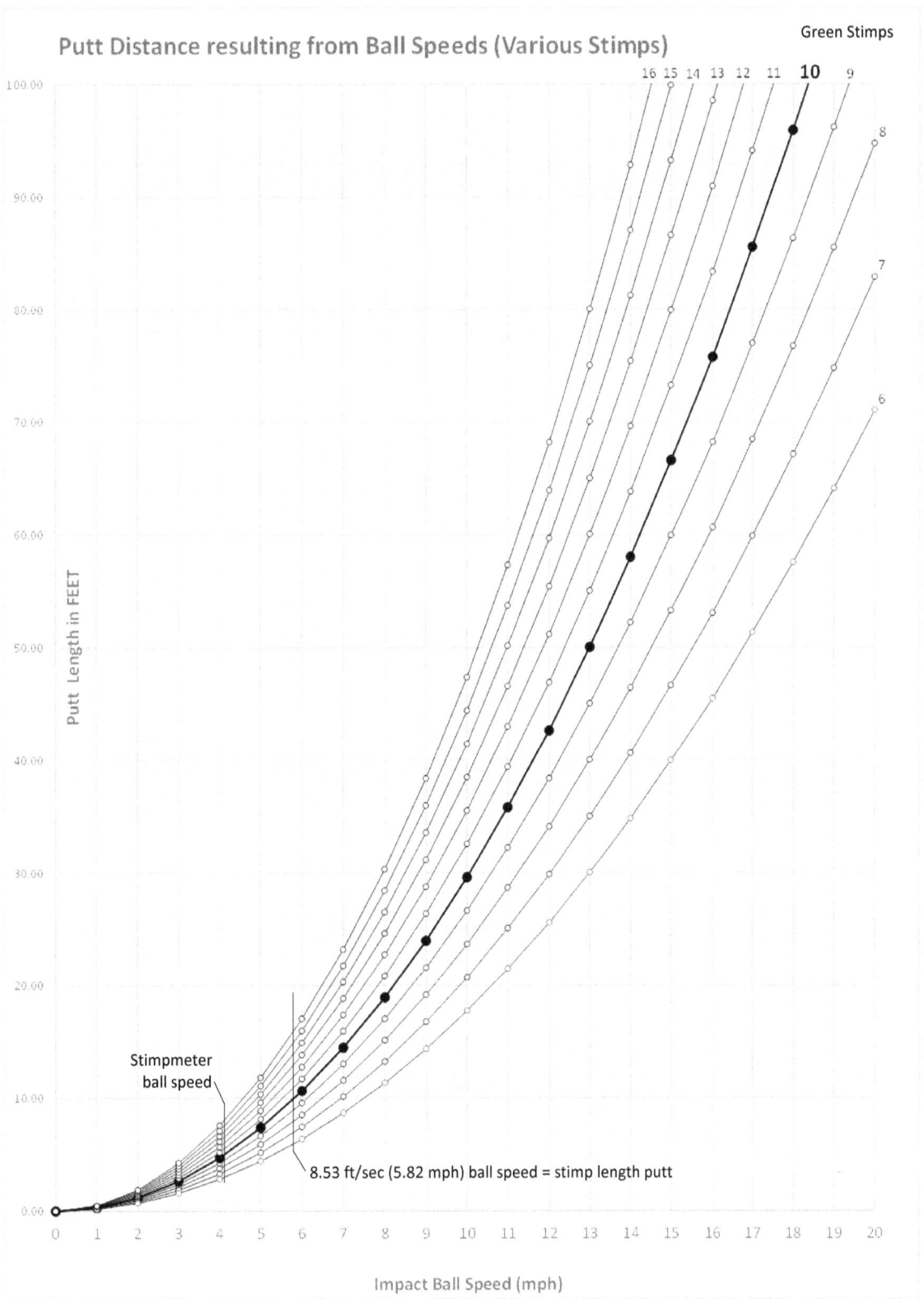

Table of Roll Distance (ft/sec ball speed)

stimps	6	7	8	9	10	11	12	13	14	15	16
0.00	0.00	0.00	0.00	0.00	0.00	0.00	0.00	0.00	0.00	0.00	0.00
1.00	0.08	0.10	0.11	0.12	0.14	0.15	0.17	0.18	0.19	0.21	0.22
2.00	0.33	0.39	0.44	0.50	0.55	0.61	0.66	0.72	0.77	0.83	0.88
3.00	0.74	0.87	0.99	1.11	1.24	1.36	1.49	1.61	1.73	1.86	1.98
4.00	1.32	1.54	1.76	1.98	2.20	2.42	2.64	2.86	3.08	3.30	3.52
5.00	2.06	2.41	2.75	3.09	3.44	3.78	4.13	4.47	4.81	5.16	5.50
6.00	2.97	3.47	3.96	4.46	4.95	5.45	5.94	6.44	6.93	7.43	7.92
7.00	4.04	4.72	5.39	6.07	6.74	7.41	8.09	8.76	9.44	10.11	10.78
8.00	5.28	6.16	7.04	7.92	8.80	9.68	10.56	11.44	12.32	13.20	14.08
9.00	6.68	7.80	8.91	10.03	11.14	12.26	13.37	14.48	15.60	16.71	17.83
10.00	8.25	9.63	11.00	12.38	13.75	15.13	16.51	17.88	19.26	20.63	22.01
11.00	9.99	11.65	13.31	14.98	16.64	18.31	19.97	21.64	23.30	24.96	26.63
12.00	11.88	13.86	15.84	17.83	19.81	21.79	23.77	25.75	27.73	29.71	31.69
13.00	13.95	16.27	18.60	20.92	23.24	25.57	27.89	30.22	32.54	34.87	37.19
14.00	16.18	18.87	21.57	24.26	26.96	29.65	32.35	35.05	37.74	40.44	43.13
15.00	18.57	21.66	24.76	27.85	30.95	34.04	37.14	40.23	43.33	46.42	49.52
16.00	21.13	24.65	28.17	31.69	35.21	38.73	42.25	45.77	49.30	52.82	56.34
17.00	23.85	27.82	31.80	35.77	39.75	43.72	47.70	51.67	55.65	59.62	63.60
18.00	26.74	31.19	35.65	40.11	44.56	49.02	53.48	57.93	62.39	66.85	71.30
19.00	29.79	34.76	39.72	44.69	49.65	54.62	59.58	64.55	69.51	74.48	79.44
20.00	33.01	38.51	44.01	49.52	55.02	60.52	66.02	71.52	77.02	82.53	88.03
21.00	36.39	42.46	48.53	54.59	60.66	66.72	72.79	78.85	84.92	90.98	97.05
22.00	39.94	46.60	53.26	59.91	66.57	73.23	79.88	86.54	93.20	99.86	106.51
23.00	43.66	50.93	58.21	65.48	72.76	80.04	87.31	94.59	101.86	109.14	116.42
24.00	47.53	55.46	63.38	71.30	79.22	87.15	95.07	102.99	110.91	118.84	126.76
25.00	51.58	60.18	68.77	77.37	85.96	94.56	103.16	111.75	120.35	128.95	137.54
26.00	55.79	65.09	74.38	83.68	92.98	102.28	111.57	120.87	130.17	139.47	148.77
27.00	60.16	70.19	80.22	90.24	100.27	110.30	120.32	130.35	140.38	150.40	160.43
28.00	64.70	75.48	86.27	97.05	107.83	118.62	129.40	140.18	150.97	161.75	172.53
29.00	69.40	80.97	92.54	104.11	115.67	127.24	138.81	150.38	161.94	173.51	185.08
30.00	74.27	86.65	99.03	111.41	123.79	136.17	148.55	160.93	173.30	185.68	198.06

Ball Speed at Impact (ft/sec)

Ball Speed in ft/sec

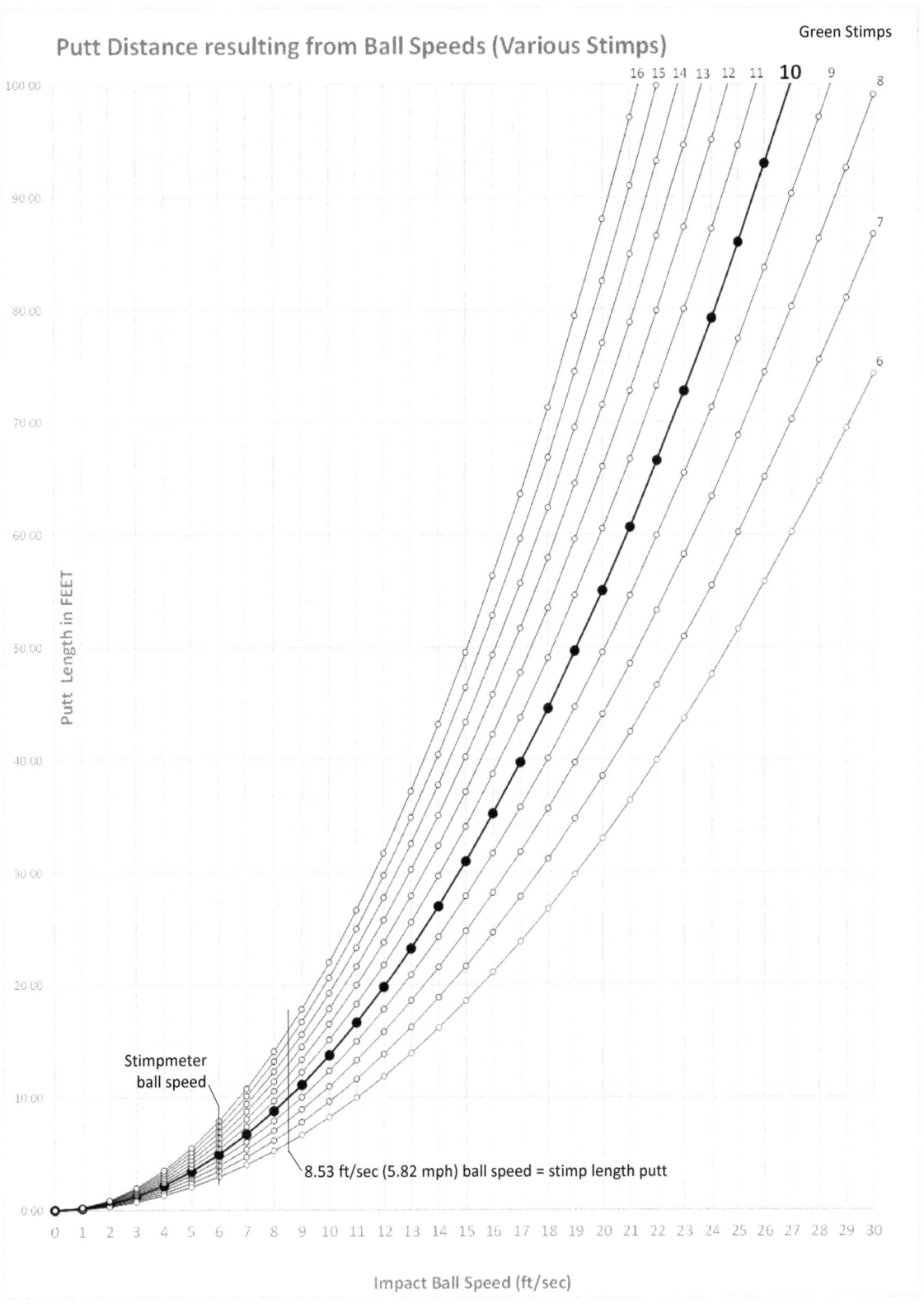

Ball Roll Physics
2.4 Impact Ratio

Impact Ratio is smash factor for putting, ball speed divided by putter speed. Impact ratio is crucial in putting physics because it determines ball speed. The putt distance tables and graphs immediately above are based on ball speed. A higher impact ratio will increase ball speed, obviously. Tests with humans putting have higher impact ratio than robot tests. The reason likely relates to the inertia of the putting stroke moving body parts.

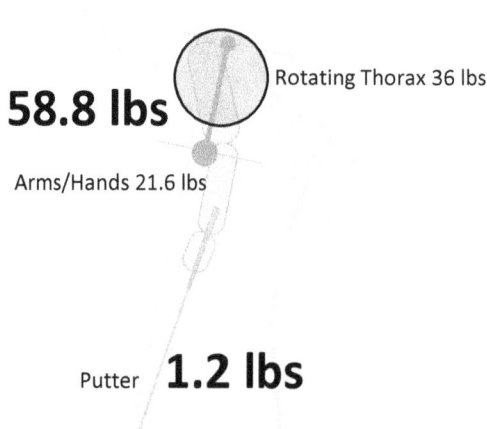

Body Weight Segment Inertia Visualized

The inertia of the moving body parts in the putting stroke improves the efficiency of the energy transfer between the putter and ball at impact. A second characteristic of human putting is acceleration at impact, which has been shown to increase impact ratio also.

Rod Spittle, a large, tall, strong man, with an accelerating putting stroke had an impact ratio of 1.85 with a putter that tested on a robot at 1.75.

The average impact ratio for most players is likely about 1.75 with a robot tested impact ratio of 1.70.

A very stiff shaft can also increase the efficiency of energy transfer, plus limiting oscillations of the shaft. A very stiff shaft can increase impact ratio from 1.70 to 1.74, not insignificant.

A very soft face insert will reduce impact ratio from 1.70 to about 1.65, not insignificant.

Even grip pressure can affect impact ratio.

All players should know their personal putting impact ratio.

See Section 9 Putting Launch Monitor Metrics for a discussion of the available putting launch monitors that provide impact ratio information.

Section 3
Backstroke Length Physics

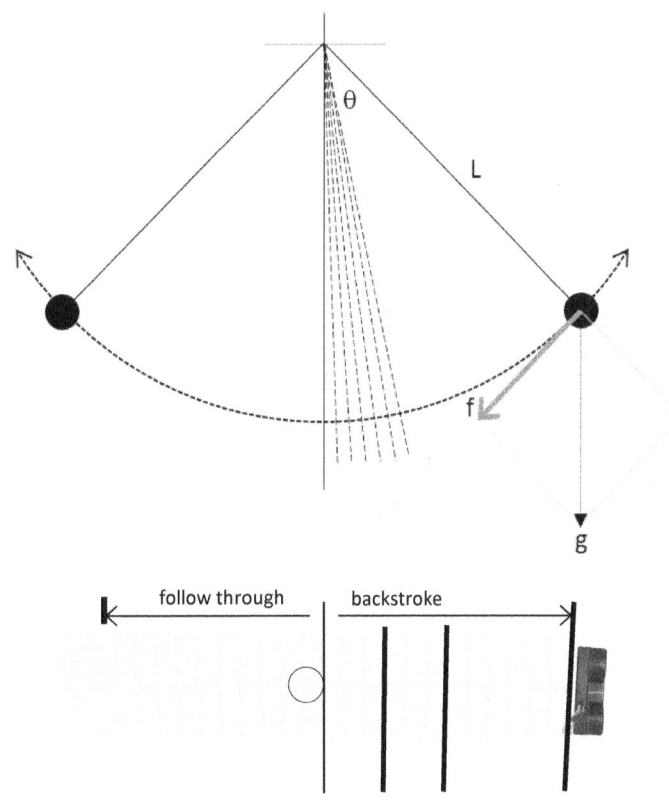

Backstroke Length
3.1 Putter Speed and Backstroke Length Physics

The putter (pendulum) moves faster as it swings further. The reason: the vector of the force of gravity acting on the pendulum in line with the swing is stronger as the putter (pendulum) swings further. See acceleration force "f" is the figures below.

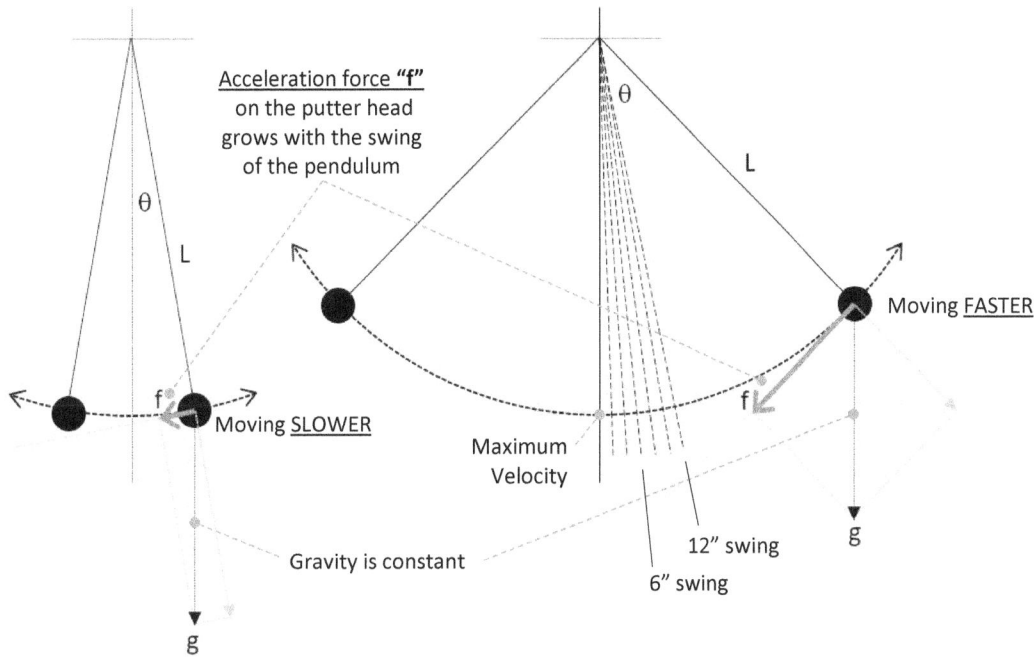

The increase in putter (pendulum) speed with stroke length (backstroke length) is VERY precise. With the constant acceleration of gravity a 12" inch swing produces twice the maximum velocity (putter speed) of a 6" swing. Every increase in swing increases the speed of the pendulum.

Every added inch of pendulum swing adds same additional pendulum speed.

That is, if 6" backstroke produces a 2 mph putter speed, a 12" backstroke will produce a 4 mph putter speed; and a 24" backstroke will produce an 8 mph putter speed.

This characteristic of theoretical pendulums is true in the modern pendulum tempo putting stroke with constant acceleration; and remains true throughout the relatively small angles of the putting stroke; and remains generally true even at the larger angles required for longer putts.

These pendulum physics principles allowed us to focus on Putter mph per Backstroke Inch (Pmph/BS") as a new and revolutionary putting launch monitor metric. You have likely never heard of it; but over the next few years it will change the way putting is taught and coached; and it will change the way you approach putting.

Putter mph per Backstroke Inch (Pmph/BS")

Putter mph per Backstroke Inch (Pmph/BS") ranges from about 0.30 Pmph/BS" to 0.50 Pmph/BS". The average is about 0.36 Pmph/BS".

Pendulums are amazing, and regular (consistent) in many ways that are important to putting.

Putter mph per Backstroke Inch (Pmph/BS") – An Example

0.36 Pmph/BS" is shown in the example below for simplicity. The impact ratio is assumed to be 1.70.

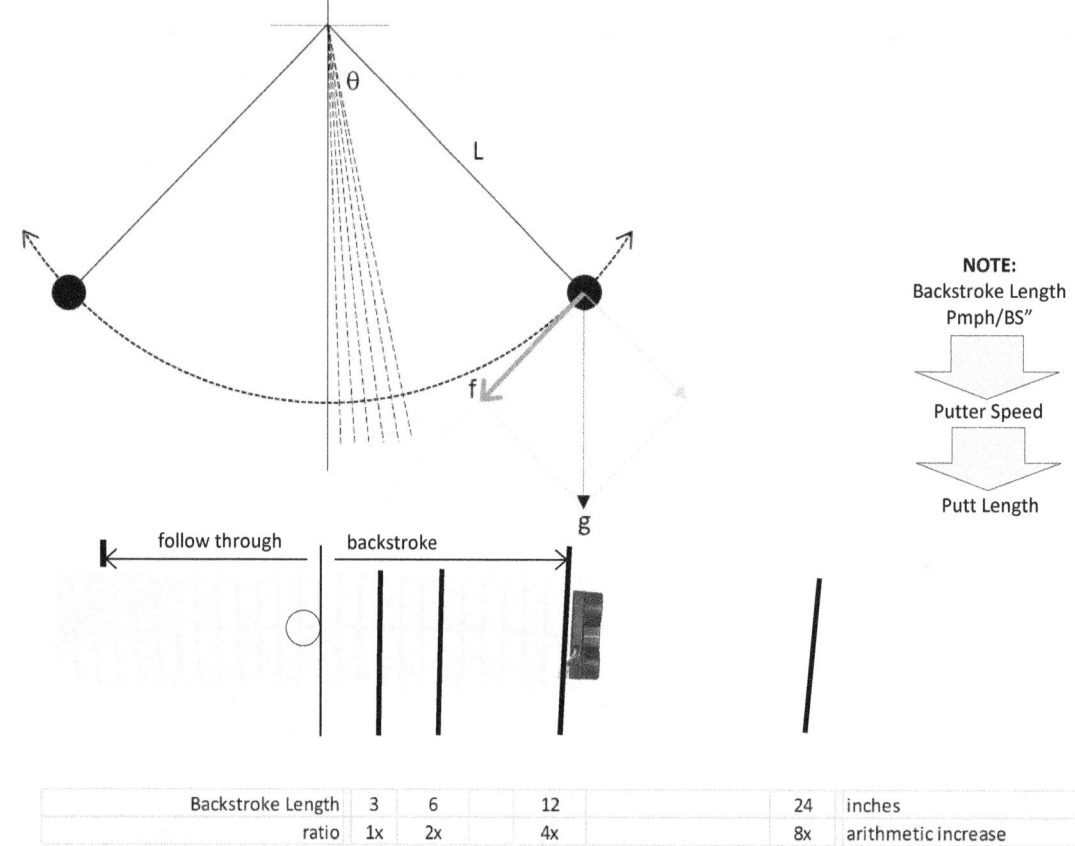

Backstroke Length	3	6	12	24	inches
ratio	1x	2x	4x	8x	arithmetic increase

A 3.00" backstroke (at 0.36 Pmph/BS") produces a 1.08 mph putter speed. 6",12" and 24" backstrokes similarly produce putter speeds of 2.16, 4.32 and 8.64 mph. A linear arithmetic relationship.

Putter Speed at Impact	1.08	2.16	4.32	8.64	mph
ratio	1x	2x	4x	8x	arithmetic increase

Assuming a constant 1.70 impact ratio, the ball speed is always 1.70 times the putter speed.

Ball Speed at impact	1.84	3.67	7.34	14.69	mph with 1.70 impact ratio
ratio	1x	2x	4x	8x	arithmetic increase

Note: 1.84 mph ball speed = 1.0' putt length. Doubling the ball speed to 3.67 mph increases the putt length to 4' - a geometric increase (4x putt length for 2x ball speed). Doubling the ball speed again to 7.34 mph increases the putt length to 16', and so on.

Putt Length on 10 stimp green	1	4	16	64	feet
ratio	1x	4x	16x	64x	geometric increase

You can see these relationships on the graph of putt distance (ball speed in mph) above on page 31.

Backstroke length determines putter speed, impact ratio then determines ball speed; and ball speed determines putt distance. This is the physics of the modern pendulum putting stroke and ball roll.

Backstroke Length

3.2 Measuring Putter mph per Backstroke Inch

Putter speed at impact and backstroke length can be measured on a putting launch monitor. Putter mph divided by backstroke length in inches equals putter mph per backstroke inch (Pmph/BS").

See Section 9 Putting Launch Monitor Metrics for a discussion of the available putting launch monitors that provide backstroke length and ball speed information.

Blast Motion data here indicates 0.33 Pmph/BS" (3.8 mph impact stroke speed/11.5" backstroke length)

$$\frac{3.8 \text{ mph impact putter speed}}{11.5\text{" backstroke length}} = 0.33 \text{ Pmph/BS"}$$

Not all putting launch monitors provide backstroke length. Trackman provides it, but rounded to full inches (not very useful, they will fix this is the future). This arcane metric will prove to be very important. You may have never heard of it. In the near future it will become widely known and related to the pendulum putting stroke and pendulum physics.

Average Putter mph per Backstroke Inch of PGA Tour Players

The SAM PuttLab data base of 99 PGA Tour players included backstroke length and putter speed information (see table below). The SAM PuttLab: Concept and PGA Tour Data for 99 PGA Tour Players. The backstroke length and putter speed data is summarized in the table below.

SAM PuttLab PGA Tour Data Analysis
Data of 99 PGA Tour players at 9 PGA tournaments between 2003 and 2005
For a straight and level 4m (13.1 ft) putt (on slower greens the distance was slightly reduced)
Reference target putter speed at impact was 1.5m/s (3.36 mph)

	normal distribution							Group SD	Individual SD
	-3SD	-2SD	-1SD	Group Average	+1SD	+2SD	+3SD		
Backswing Time (BST) seconds	0.40	0.49	0.58	0.67	0.76	0.85	0.94	0.09	0.03
			Tiger Woods	0.660			0.977	Loren Roberts	
Backswing Length (BSL) inches	5.00	6.50	7.99	9.49	10.98	12.48	13.98	1.50	0.39
Impact Speed (BSL) mph	2.58	2.85	3.11	3.38	3.64	3.91	4.18	0.27	0.10
→ Putter mph per Backstroke Inch (Pmph/BS")	0.52	0.44	0.39	0.36	0.33	0.31	0.30		

Putter mph per Backstroke Inch (Pmph/BS") is a very durable and consistent characteristic of the modern putting stroke and of pendulum physics; remaining constant over a wide range of backstroke lengths and pendulum swings. The principle allows the calculation of backstroke length required to deliver the putter at the speed required for a putt of any length. Backstroke length directly controls putt length. Tour players have a consistent pendulum tempo and vary their backstroke length to control distance. They just feel it; but you CAN know it. Very precisely.

Backstroke Length
3.3 Personal Putting Stroke Energy Varies

In the graph below the quality of release at impact ranges from line 1 (dead hands) to line 5 (strong release). Faster tempo plus stronger release at impact equals more energetic stroke. Slower tempo with less release at impact equals a less energetic stroke. Most players will be in the shaded area.

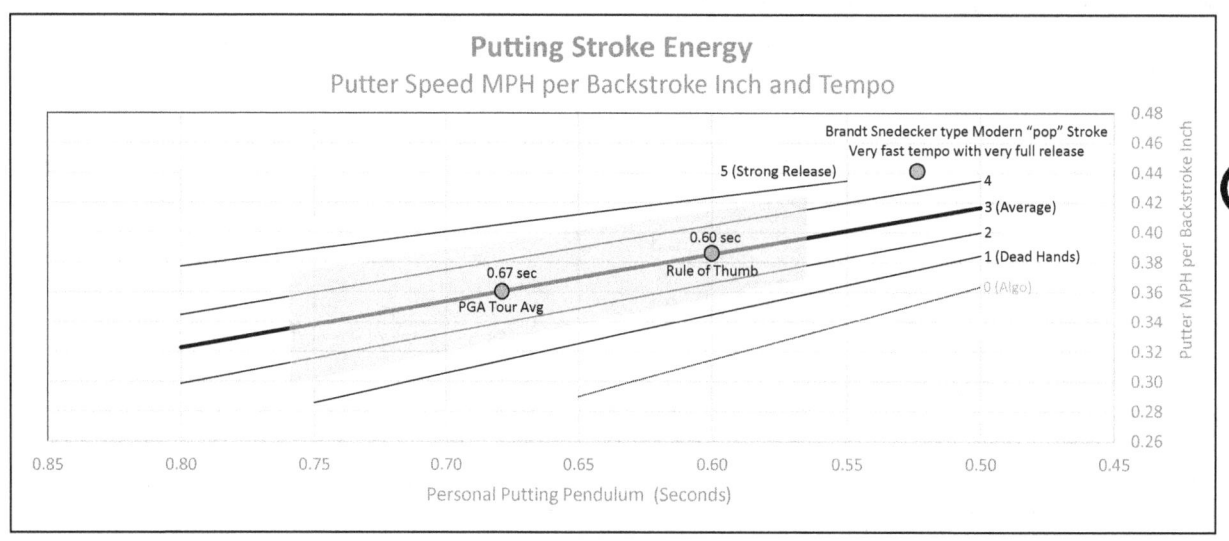

The 0 (Algo) line on the graph above is the theoretical (weightless string zero friction pivot) pendulum second harmonic. The actual human results range from lines 1 (dead hands) to line 5 (strong release). The two axes in the chart above are illustrated below:

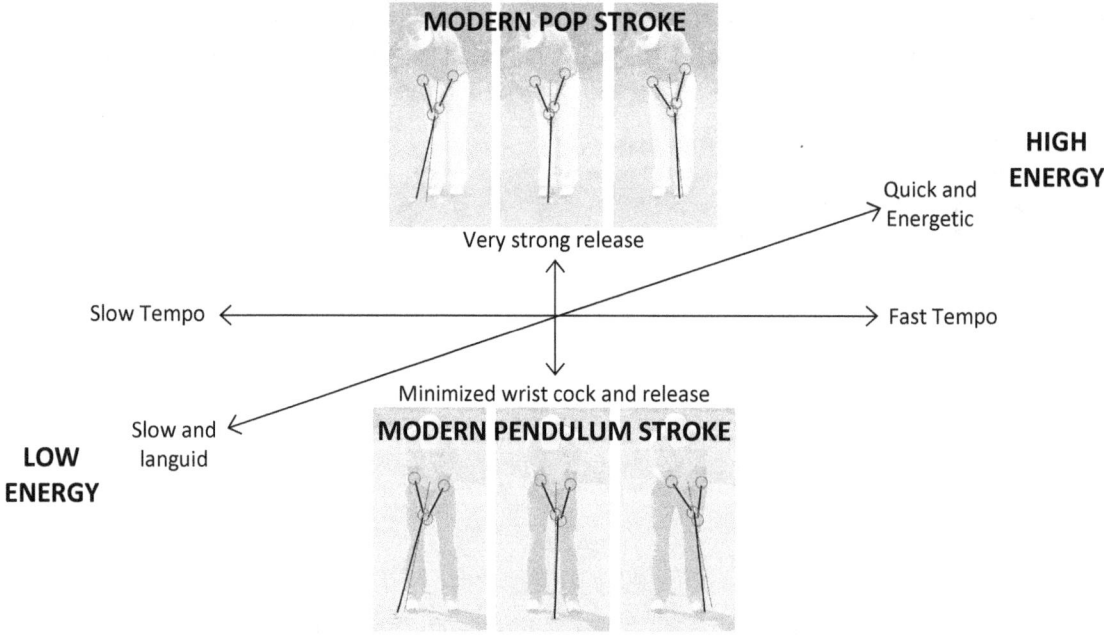

Your personal Putter mph per Backstroke Inch (Pmph/BS") will vary based on the energy of your personal putting stroke and the quality of "release at impact" in your putting stroke. A very energetic stroke (like Brandt Snedecker with a little pop at impact) will have more mph per backstroke inch than a more languid modern pendulum putting stroke.

There is no right answer; only your personal Pmph/BS".

The table below shows variation related to Pmph/BS" on a 10 stimp green with 1.75 impact ratio.

Your backstroke length with a modern pendulum tempo putting stroke, with constant acceleration, will look something like the backstroke lengths shown below depending on your personal Pmph/BS" and the speed of the green. All values are for a level putt.

Pmph/BS" varies from about 0.30 mph for the most languid stroke of very tall dead hands players to 0.50 mph for the energetic stroke of very short players with a very strong release at impact. The average is likely about 0.36 Pmph/BS".

	Putting Stroke Energy					
1.75 Impact Ratio			Pmph/BS" on a 10 stimp Green			
	0.32	0.34	0.36	0.38	0.40	0.42
3'	6.3	5.9	5.6	5.3	5.0	4.8
6'	8.5	8.0	7.5	7.1	6.8	6.5
9'	10.2	9.6	9.1	8.6	8.2	7.8
12'	11.7	11.0	10.4	9.8	9.3	8.9
15'	13.0	12.2	11.5	10.9	10.4	9.9
18'	14.2	13.3	12.6	11.9	11.3	10.8
21'	15.3	14.4	13.6	12.9	12.2	11.6
24'	16.3	15.3	14.5	13.7	13.0	12.4
27'	17.3	16.2	15.3	14.5	13.8	13.2
30'	18.2	17.1	16.2	15.3	14.5	13.8
40'	20.9	19.7	18.6	17.6	16.7	15.9
50'	23.4	22.0	20.8	19.7	18.7	17.8
60'	25.6	24.1	22.7	21.5	20.4	19.5
70'	27.6	26.0	24.5	23.2	22.1	21.0
80'	29.5	27.7	26.2	24.8	23.6	22.5
90'	31.2	29.4	27.8	26.3	25.0	23.8
100'	32.9	31.0	29.3	27.7	26.3	25.1

You may be anywhere between 0.30 Pmph/BS" and 0.50 Pmph/BS". The 0.36 Pmph/BS" is highlighted because it is about average, and a good place to start if you don't know your personal Pmph/BS".

You can see in the table above that on a 10 stimp green the backstroke required for a 12' putt is likely to be within a couple of inches of 11.1" no matter how energetic or languid your putting stroke is.

But the BIG picture is that with a modern pendulum tempo putting stroke:

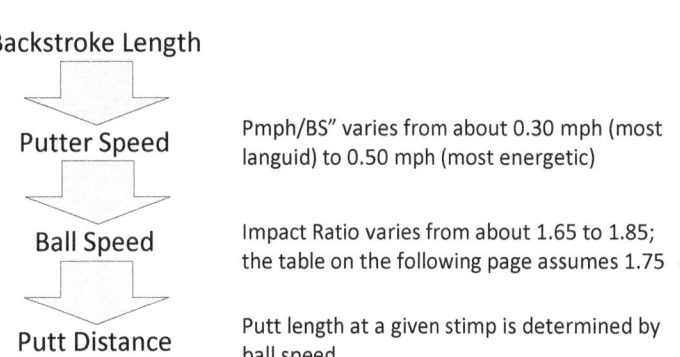

Backstroke Length
⬇
Putter Speed — Pmph/BS" varies from about 0.30 mph (most languid) to 0.50 mph (most energetic)
⬇
Ball Speed — Impact Ratio varies from about 1.65 to 1.85; the table on the following page assumes 1.75
⬇
Putt Distance — Putt length at a given stimp is determined by ball speed

Putter mph per backstroke inch is regularized by the pendulum tempo in the modern putting stroke. Your practice should always focus on tempo. Your personal pendulum tempo developed by constant acceleration in the modern pendulum tempo putting stroke allows a new focus on tempo and backstroke length.

Backstroke Length

3.4 Backstroke Length for Putts on Level Greens

Backstroke Length for 0.36 Pmph/BS" putting stroke on varying stimp Greens

Pmph/BS" Inches past hole	0.36	4	5	6	7	8	9	10	11	12	13	14	15	16	17	18	19	20	25	30	40	50	1.75 Impact Ratio
	8 Inches past hole																						
3	8.8	7.9	7.2	6.7	6.2	5.9	5.6	5.3	5.1	4.9	4.7	4.6	4.4	4.3	4.2	4.1	3.9	3.5	3.2	2.8	2.5	3
6	11.9	10.7	9.7	9.0	8.4	7.9	7.5	7.2	6.9	6.6	6.4	6.1	6.0	5.8	5.6	5.5	5.3	4.8	4.3	3.8	3.4	6
9	14.3	12.8	11.7	10.8	10.1	9.6	9.1	8.6	8.3	8.0	7.7	7.4	7.2	7.0	6.8	6.6	6.4	5.7	5.2	4.5	4.1	9
12	16.4	14.7	13.4	12.4	11.6	10.9	10.4	9.9	9.5	9.1	8.8	8.5	8.2	8.0	7.7	7.5	7.3	6.6	6.0	5.2	4.6	12
15	18.3	16.3	14.9	13.8	12.9	12.2	11.5	11.0	10.5	10.1	9.8	9.4	9.1	8.9	8.6	8.4	8.2	7.3	6.7	5.8	5.2	15
18	19.9	17.8	16.3	15.1	14.1	13.3	12.6	12.0	11.5	11.1	10.7	10.3	10.0	9.7	9.4	9.1	8.9	8.0	7.3	6.3	5.6	18
21	21.5	19.2	17.5	16.2	15.2	14.3	13.6	12.9	12.4	11.9	11.5	11.1	10.7	10.4	10.1	9.9	9.6	8.6	7.8	6.8	6.1	21
24	22.9	20.5	18.7	17.3	16.2	15.3	14.5	13.8	13.2	12.7	12.2	11.8	11.5	11.1	10.8	10.5	10.2	9.2	8.4	7.2	6.5	24
27	24.3	21.7	19.8	18.3	17.2	16.2	15.3	14.6	14.0	13.5	13.0	12.5	12.1	11.8	11.4	11.1	10.8	9.7	8.9	7.7	6.9	27
30	25.5	22.8	20.9	19.3	18.1	17.0	16.2	15.4	14.7	14.2	13.7	13.2	12.8	12.4	12.0	11.7	11.4	10.2	9.3	8.1	7.2	30
40	29.4	26.3	24.0	22.2	20.8	19.6	18.6	17.7	17.0	16.3	15.7	15.2	14.7	14.3	13.9	13.5	13.2	11.8	10.7	9.3	8.3	40
50	32.8	29.4	26.8	24.8	23.2	21.9	20.8	19.8	19.0	18.2	17.5	17.0	16.4	15.9	15.5	15.1	14.7	13.1	12.0	10.4	9.3	50
60	35.9	32.1	29.3	27.2	25.4	23.9	22.7	21.7	20.7	19.9	19.2	18.6	18.0	17.4	16.9	16.5	16.1	14.4	13.1	11.4	10.2	60
70	38.8	34.7	31.7	29.3	27.4	25.8	24.5	23.4	22.4	21.5	20.7	20.0	19.4	18.8	18.3	17.8	17.3	15.5	14.2	12.3	11.0	70
80	41.4	37.0	33.8	31.3	29.3	27.6	26.2	25.0	23.9	23.0	22.1	21.4	20.7	20.1	19.5	19.0	18.5	16.6	15.1	13.1	11.7	80
90	43.9	39.3	35.9	33.2	31.1	29.3	27.8	26.5	25.4	24.4	23.5	22.7	22.0	21.3	20.7	20.1	19.6	17.6	16.0	13.9	12.4	90
100	46.3	41.4	37.8	35.0	32.7	30.8	29.3	27.9	26.7	25.7	24.7	23.9	23.1	22.4	21.8	21.2	20.7	18.5	16.9	14.6	13.1	100

Putter mph per Backstroke Inch (Pmph/BS" – putting stroke energy – 0.36 Pmph/BS" on the card above), with a given impact ratio (1.75 on the card above) provides a very specific backstroke length for level putts of various length on greens of various stimps.

Pendulum Putting Tempo Practice Drill on a Putting Green

Find a level place to practice your backstroke length on putts of various lengths. A simple practice drill can combine tempo and backstroke length. Practice a series of putts: 3', 6' and 12' every time you practice putting, confirming your personal pendulum putting tempo in your muscle memory. Sometimes add a 24' , 50' and 100' putts.

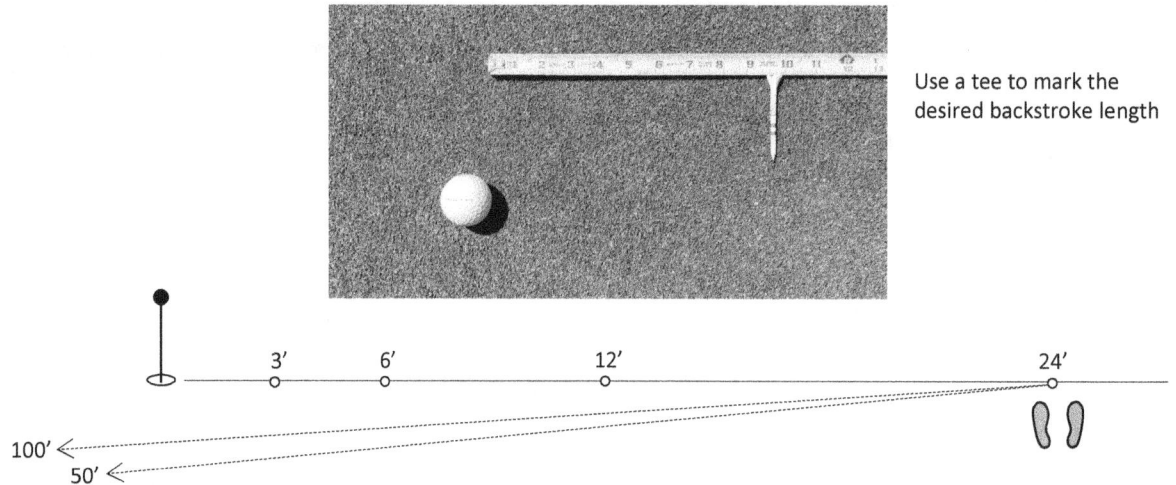

Use a tee to mark the desired backstroke length

You can use your own sense of timing by silently singing a song phrase or counting under your breath, 1,2,3, with a strong simple rhythm. Whatever you use, metronome, song or counting, you need to fit the 3', 6', 12', 24', 50' and 100' putts into your tempo, varying only the backstroke length and providing the constant acceleration required to get the putter moving faster and faster in the longer and longer putts.

The goal is to confirm your personal pendulum putting tempo and backstroke length for these putts by practicing on the green.

Section 4
Sloping Greens Backstroke Length Physics

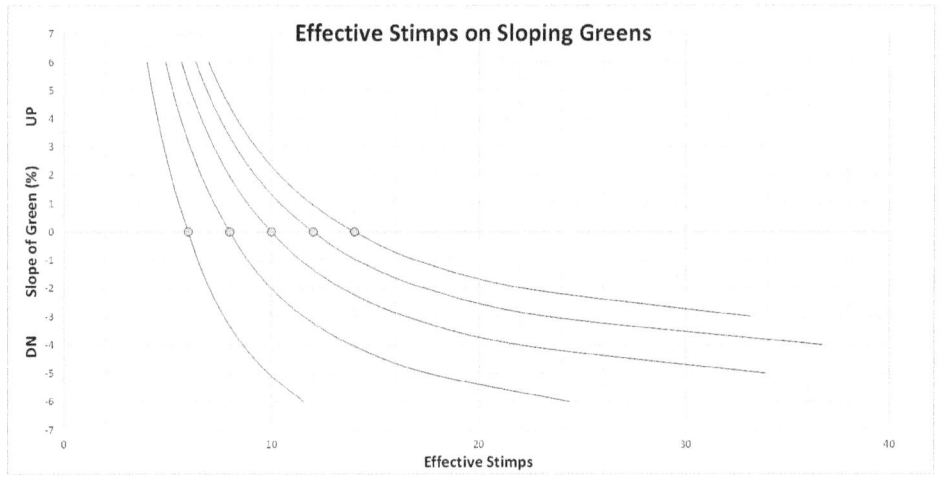

Sloping Greens Backstroke Length Physics
4.1 Effective Stimps for Sloping Greens

The effective stimp of a sloping green is the level green stimp that will roll the same distance.

A 2% downhill putt on a 10 stimp green will roll like a level putt on a 13.4 stimp green. A 2% uphill putt on a 10 stimp green will roll like a level putt on an 8 stimp green. Knowing the effective stimp of a putt will allow you to use the level green backstroke length to control putt distance on sloping greens. The effective stimp for sloping greens in the table below has been established by calculation with a complex algorithm and experimentation with a stimpmeter on greens of known slopes.

stimp	6	7	8	9	10	11	12	13	14
up 6%	4.02	4.48	4.91	5.31	5.67	6.03	6.36	6.67	6.97
up 5%	4.26	4.77	5.25	5.70	6.11	6.51	6.89	7.24	7.59
up 4%	4.53	5.10	5.64	6.15	6.62	7.08	7.52	7.93	8.33
up 3%	4.83	5.47	6.09	6.67	7.23	7.76	8.28	8.76	9.24
up 2%	5.16	5.90	6.61	7.30	7.95	8.59	9.21	9.80	10.38
up 1%	5.55	6.40	7.24	8.06	8.85	9.64	10.41	11.15	11.90
level	6.00	7.00	8.00	9.00	10.00	11.00	12.00	13.00	14.00
down 1%	6.45	7.63	8.85	10.10	11.39	12.72	14.10	15.51	16.98
down 2%	7.01	8.44	9.98	11.63	13.40	15.32	17.41	19.67	22.16
down 3%	7.71	9.51	11.55	13.87	16.52	19.63	23.31	27.73	33.18
down 4%	8.63	10.99	13.86	17.44	22.01	28.13	36.74	49.79	72.10
down 5%	9.85	13.12	17.56	23.93	33.93	52.00	94.97		
down 6%	11.55	16.47	24.40	39.42	79.28				

You can see in the table above and graph below that the adjustment is more dramatic for downhill putts. The Δ dn is much larger than the Δ up. Downhill putts are much trickier than uphill putts.

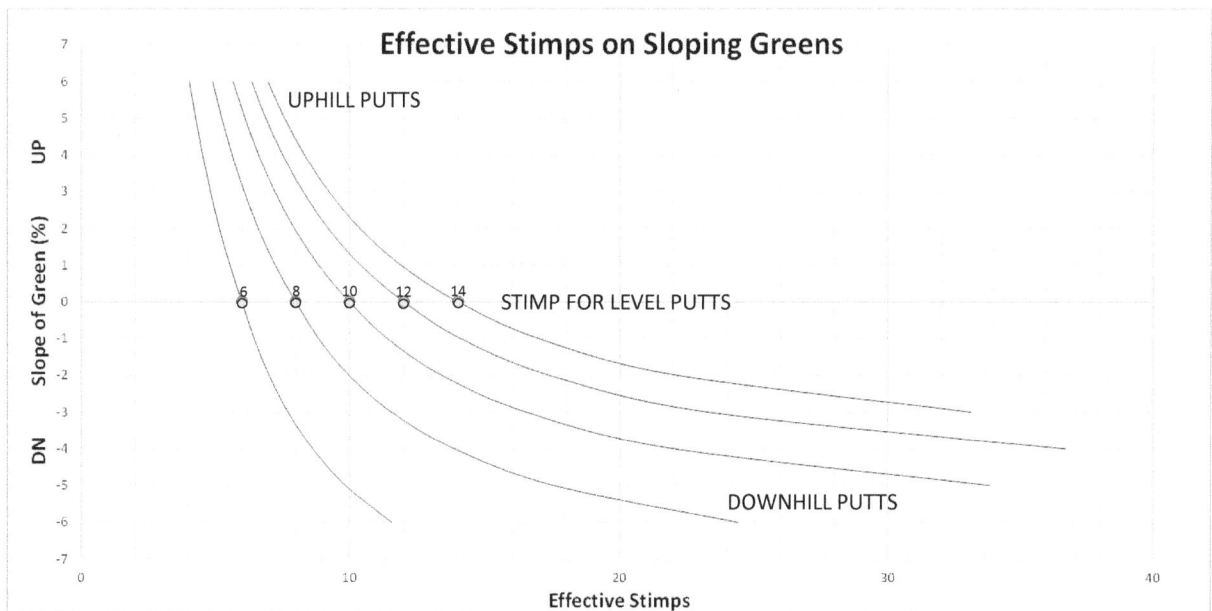

The adjustment required for uphill putts is small compared to downhill putts. You can SEE in the graph above why severe downhill putts are feared by players of all calibers. A 3% downhill putt at Oakmont or at the Masters on a 14 stimp green has an effective stimp of about 33.2! A 4% downhill putt has an effective stimp of 72.1. A ball will not stop on a 5% downhill slope on a 14 stimp green. In fact it will accelerate.

Sloping Greens Backstroke Length Physics
4.2 Estimating the % Slope of Greens

Aimpoint trained players are used to estimating the % slope of greens visually and by feeling it with their feet. We have used % slope because both the USGA and Aimpoint use it. You can make a useful guess about green slopes with the characteristics of various percent slopes on 10 stimp greens noted below:

barely discernable, very minimal slope
1% — 0:100 / 1:100

clearly visible slope
2% — 0:100 / 2:100

clearly visible strong slope; likely too much slope for a hole location
3% — 0:100 / 3:100

clearly too much slope for a hole location
4% — 0:100 / 4:100

ball will not come to rest rolling downhill
5% — 0:100 / 5:100

ball will come back to your feet rolling uphill
6% — 0:100 / 6:100

Most digital levels can display results in either % slope or degrees. The "Breakmaster" (right below) displays only in degrees; but shows the fall line with an arrow. The table below will allow conversion between the units.

% slope	degrees	rounded degrees	rule of thumb	degrees	% slope
6	3.434	3.4	3.6	6	10.510
5	2.862	2.9	3.0	5	8.749
4	2.291	2.3	2.4	4	6.993
3	1.718	1.7	1.8	3	5.241
2	1.146	1.1	1.2	2	3.492
1	0.573	0.6	0.6	1	1.746
0	0.000	0.0	0.0	0	0.000

Rule of Thumb: 1% slope = 0.6 degrees

2.7 degrees = 4.7% slope

Sloping Greens Backstroke Length Physics

4.3 Backstroke Length Card for Sloping Green

Determine the slope of the green/putt and its effective stimp (top table below). Find the required backstroke length for a level putt on a green of the effective stimp in the second table. Every putt is effectively converted to a putt on a level green.

Effective stimps on 6 to 14 stimp sloping greens

stimp	6	7	8	9	10	11	12	13	14
up 6%	4.02	4.48	4.91	5.31	5.67	6.03	6.36	6.67	6.97
up 5%	4.26	4.77	5.25	5.70	6.11	6.51	6.89	7.24	7.59
up 4%	4.53	5.10	5.64	6.15	6.62	7.08	7.52	7.93	8.33
up 3%	4.83	5.47	6.09	6.67	7.23	7.76	8.28	8.76	9.24
up 2%	5.16	5.90	6.61	7.30	7.95	8.59	9.21	9.80	10.38
up 1%	5.55	6.40	7.24	8.06	8.85	9.64	10.41	11.15	11.90
level	6.00	7.00	8.00	9.00	10.00	11.00	12.00	13.00	14.00
down 1%	6.45	7.63	8.85	10.10	11.39	12.72	14.10	15.51	16.98
down 2%	7.01	8.44	9.98	11.63	13.40	15.32	17.41	19.67	22.16
down 3%	7.71	9.51	11.55	13.87	16.52	19.63	23.31	27.73	33.18
down 4%	8.63	10.99	13.86	17.44	22.01	28.13	36.74	49.79	72.10
down 5%	9.85	13.12	17.56	23.93	33.93	52.00	94.97		
down 6%	11.55	16.47	24.40	39.42	79.28				

0.36	4	5	6	7	8	9	10	11	12	13	14	15	16	17	18	19	20	25	30	40	50	1.75
8 inches past hole																						
3	8.8	7.9	7.2	6.7	6.2	5.9	5.6	5.3	5.1	4.9	4.7	4.6	4.4	4.3	4.2	4.1	3.9	3.5	3.2	2.8	2.5	3
6	11.9	10.7	9.7	9.0	8.4	7.9	7.5	7.2	6.9	6.6	6.4	6.1	6.0	5.8	5.6	5.5	5.3	4.8	4.3	3.8	3.4	6
9	14.3	12.8	11.7	10.8	10.1	9.6	9.1	8.6	8.3	8.0	7.7	7.4	7.2	7.0	6.8	6.6	6.4	5.7	5.2	4.5	4.1	9
12	16.4	14.7	13.4	12.4	11.6	10.9	10.4	9.9	9.5	9.1	8.8	8.5	8.2	8.0	7.7	7.5	7.3	6.6	6.0	5.2	4.6	12
15	18.3	16.3	14.9	13.8	12.9	12.2	11.5	11.0	10.5	10.1	9.8	9.4	9.1	8.9	8.6	8.4	8.2	7.3	6.7	5.8	5.2	15
18	19.9	17.8	16.3	15.1	14.1	13.3	12.6	12.0	11.5	11.1	10.7	10.3	10.0	9.7	9.4	9.1	8.9	8.0	7.3	6.3	5.6	18
21	21.5	19.2	17.5	16.2	15.2	14.3	13.6	12.9	12.4	11.9	11.5	11.1	10.7	10.4	10.1	9.9	9.6	8.6	7.8	6.8	6.1	21
24	22.9	20.5	18.7	17.3	16.2	15.3	14.5	13.8	13.2	12.7	12.2	11.8	11.5	11.1	10.8	10.5	10.2	9.2	8.4	7.2	6.5	24
27	24.3	21.7	19.8	18.3	17.2	16.2	15.3	14.6	14.0	13.5	13.0	12.5	12.1	11.8	11.4	11.1	10.8	9.7	8.9	7.7	6.9	27
30	25.5	22.8	20.9	19.3	18.1	17.0	16.2	15.4	14.7	14.2	13.7	13.2	12.8	12.4	12.0	11.7	11.4	10.2	9.3	8.1	7.2	30
40	29.4	26.3	24.0	22.2	20.8	19.6	18.6	17.7	17.0	16.3	15.7	15.2	14.7	14.3	13.9	13.5	13.2	11.8	10.7	9.3	8.3	40
50	32.8	29.4	26.8	24.8	23.2	21.9	20.8	19.8	19.0	18.2	17.5	17.0	16.4	15.9	15.5	15.1	14.7	13.1	12.0	10.4	9.3	50
60	35.9	32.1	29.3	27.2	25.4	23.9	22.7	21.7	20.7	19.9	19.2	18.6	18.0	17.4	16.9	16.5	16.1	14.4	13.1	11.4	10.2	60
70	38.8	34.7	31.7	29.3	27.4	25.8	24.5	23.4	22.4	21.5	20.7	20.0	19.4	18.8	18.3	17.8	17.3	15.5	14.2	12.3	11.0	70
80	41.4	37.0	33.8	31.3	29.3	27.6	26.2	25.0	23.9	23.0	22.1	21.4	20.7	20.1	19.5	19.0	18.5	16.6	15.1	13.1	11.7	80
90	43.9	39.3	35.9	33.2	31.1	29.3	27.8	26.5	25.4	24.4	23.5	22.7	22.0	21.3	20.7	20.1	19.6	17.6	16.0	13.9	12.4	90
100	46.3	41.4	37.8	35.0	32.7	30.8	29.3	27.9	26.7	25.7	24.7	23.9	23.1	22.4	21.8	21.2	20.7	18.5	16.9	14.6	13.1	100

Backstroke Length for varying effective stimp Greens

This method is remarkably accurate. Two examples are explained below:

1. 21' putt, 9 stimp green and a 3% <u>downhill</u> slope. A very <u>strongly downhill</u> putt.
   A 3% downhill putt on a 9 stimp green has an effective stimp of 13.9 – about 14 stimp.
   An 11.5" backstroke is indicated for the 21' downhill putt.

2. 21' putt, 12 stimp green and a 3% <u>uphill</u> slope. A very <u>strongly uphill</u> putt.
   A 3% uphill putt on a 12 stimp green has an effective stimp of 8.3 – about 8 stimp.
   A 15.2" backstroke is indicated for the 21' uphill putt.

   Remember, this card is for 0.36 Pmph/BS" and 1.75 impact ratio.

Making a Feel-Based Stroke

Knowing the effective stimp of sloping greens allows the adjustment of EVERY putt to a level putt on a level green of the effective stimp of sloping putt. Using the method may seem overly analytical and unattractive for feel players. But this method can improve your feel.

A simple feel improvement method is proposed. First determine the backstroke length required by the method for your putt. Take couple of practice strokes focused on the required backstroke length and note the position of the putter in relation to your feet; and the feel of rotated torso. Step into the putt focused on the feel of the "correct" backstroke length. Make a feel-based stroke.

This method will allow you to calibrate even the most extreme putts you encounter. Two examples of severe slope follow: a severe downhill putt and a severe uphill putt.

Extreme Slopes

Consider a severe downhill putt of 24' in length and about 5% downhill on a 10 stimp green. Your friends will tell you to "just touch it" or "this is going to be tough to leave short". The backstroke length chart above provides more detailed guidance. A 5% downhill putt on 10 stimp green has an effective stimp of about 34 and a 24' putt on a 34 stimp green will require about an 8" backstroke. An 8" backstroke on a level 10 stimp green would result in about a 6' putt. Knowing that about an 8" backstroke is required allows you to make calibrated practice strokes; increasing your chance of success.

Now consider a similar uphill putt of the same 24' but 5% uphill on the same 10 stimp green. Your friends might tell you "this is going to be a very slow putt" or "you have to kill this one". A 5% uphill putt on 10 stimp green has an effective stimp of about 6 and a 24' putt on a 6 stimp green will require about a 19" backstroke. A 19" backstroke on a level 10 stimp green would result in about a 40' putt. Knowing that about a 19" backstroke is required allows you to make calibrated practice strokes; increasing your chance of success.

Note

The distance control principles outlined here apply to chipping around the green. Chip shots of less than 25 yards have the same pendulum tempo (0.60 second backstroke and 0.60 second through stroke). It would take another book to expand the discussion to cover chipping. Maybe someday we will explore it in detail. For now, just know that it is worth exploring in your personal putting and chipping practice.

This method of distance control through backstroke length is made possible by establishing a regular pendulum putting tempo for all putts of all length. You can improve your distance control dramatically by practicing tempo and backstroke length together. These together offer game changing performance.

Appendix A has a range of backstroke length cards for you to use.

An Adjustment for Grain

For players in the south putting on greens with a lot of grain (nap), an additional adjustment can be made. Most agronomists agree that grass grows downhill almost all of the time. The setting sun has much less effect on grain than the slope of the green. The grain of a green can change the effective stimp of a green only a little, and affects the break of a putt mostly in the last foot of roll. The adjustment required for stimp is therefore very small. You can round backstroke length up a little going into the grain and round backstroke length down a little going down grain. The required adjustment is very small. Gravity is the force majeure.

Sloping Greens Backstroke Length Physics

4.4 Why Stimps are Exaggerated (honest mistakes)

Average uphill and downhill stimp readings exaggerates stimps

The method of averaging the uphill and downhill measured stimps is relatively accurate on slow greens; but progressively less accurate on faster greens. The uphill stimp on a 3% slope for a 14 stimp green is about 9.24. The downhill stimp on the same green is about 33.18. The average of these two measured stimps is 21.21; remember this is a 14 stimp green. The averaging method exaggerates the stimp on faster greens if the stimps are measured in an area of high slope.

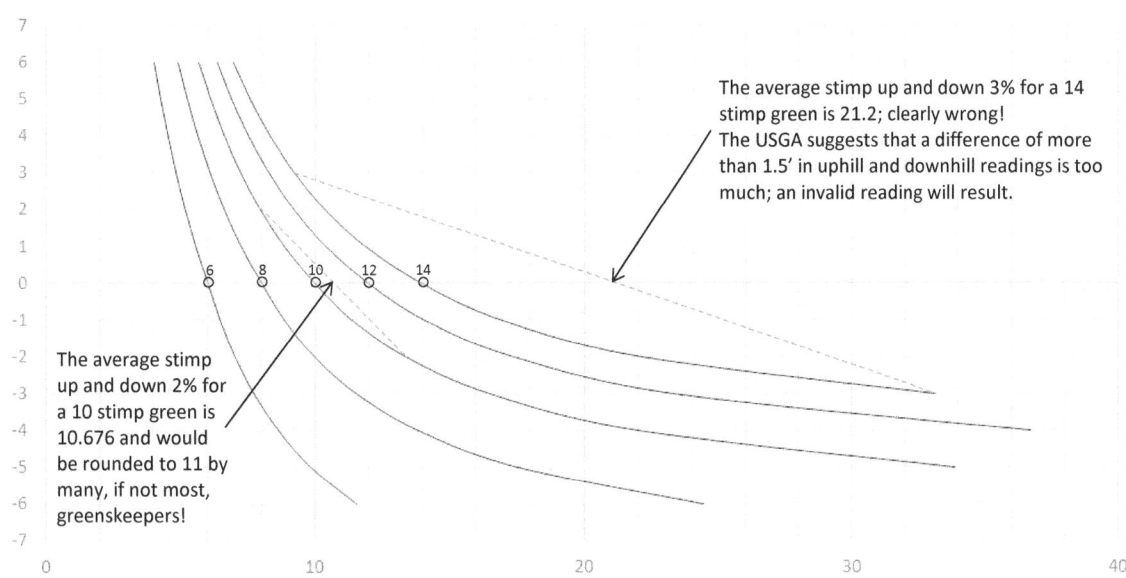

The exaggeration of slope caused by averaging uphill and downhill slopes on various stimp greens.

slope	6	8	10	12	14
3%	6.271	8.817	11.876	15.794	21.206
2%	6.088	8.295	10.676	13.309	16.270
1%	6.002	8.042	10.119	12.254	14.438
level	6.000	8.00	10.00	12.000	14.00

NOTE:
1% slopes are ok;
2% slopes work only on slow greens; above 10 stimp 2% slopes create invalid readings;
3% slope is wrong even on 8 stimp greens.

Carelessly continuing to raise the stimpmeter after the ball has left exaggerates stimps

A second cause of exaggerated stimps is from operator error with the stimpmeter. In measuring the stimp you are supposed to raise the stimpmeter to the height where the ball rolls out of its notch and down the stimpmeter. It is very common to continue raising the stimpmeter, a little, after the ball leaves its notch. This will increase the speed of the ball down the stimpmeter, as shown in the table below.

	excess	raising of stimpmeter (inches)				
ft/sec		6	8	10	12	14	stimps
6.00	0"	6.0	8.0	10.0	12.0	14.0
6.15	0.5"	6.30	8.40	10.50	12.60	14.69
6.30	1.0"	6.61	8.81	11.01	13.22	15.42

The combination of the two effects above will almost always result in an increased stimp

As an example, note that on a 10 stimp green with 2% slope:	
Averaging the uphill and downhill stimp would give a result of	10.676
Raising the stimpmeter an extra 0.5" would add	0.500
Causing a stimp result of	11.176

The correct way to determine the stimp of a sloping green is by using the Brede formula.

The Brede (A. Douglas Brede) formula allows accurate calculations: $S = 2(S_{up} \times S_{dn})/(S_{up} + S_{dn})$.
It works extremely well for slower greens, and well enough even on the fastest greens.

Brede Formula slope = $2(S_{up} \times S_{dn})/(S_{up} + S_{dn})$					
	6	8	10	12	14
3%	5.938	7.972	10.056	12.217	14.450
2%	5.948	7.953	9.981	12.048	14.140
1%	5.969	7.961	9.960	11.976	13.990
level	6.000	8.00	10.00	12.000	14.00

For the 3% slope on the 14 stimp green the calculation works like the following:
S_{up} = 9.24 S_{dn} = 33.18 $(S_{up} \times S_{dn})$ = 9.24 x 33.18 = 306.58 $(S_{up} + S_{dn})$ = 9.24 + 33.18 = 42.42.
Brede Formula Stimp = 613.16/42.42 = 14.45
which should be rounded to 14

The lesson here is that you need to avoid 3% slopes when measuring stimps; even the Brede formula begins to break at 3% slopes.

The BIG lesson here is that you need to know the actual accurately measured stimp of the greens you are putting on to use this backstroke length system effectively.

Section 5
Biomechanical Putting Posture

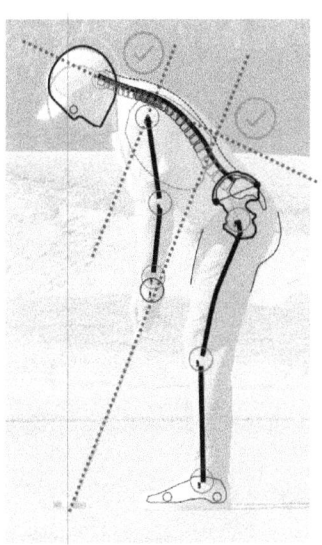

Biomechanical Putting Posture
5.1 The Two-Plane Posture Dominates the Tour

The Straight Back Straight Through (SBST) stroke, recommended by Dave Pelz, is almost mythical. The posture (right below) required for a SBST stroke is not common on tour.

The Two-Plane Neutral Arc posture (left below) dominates the PGA Tour. Why?

We believe the answer lies in the the biomechanics of spinal rotation; specifically, the freedom of rotation on the upper thoracic T3/T4 axis of rotation and the STABILITY added when the putter shaft plane of rotation is aligned with the T3/T4 axis of rotation.

We divide putting postures into 4 categories (see figures below). The 4 categories can be sorted into postures with T3/T4 aligned with the plane of the putter shaft (left two below); and postures with T3/T4 and putter shaft plane NOT aligned (two right below). The relationship of the T3/T4 axis of rotation to the putter shaft is crucial.

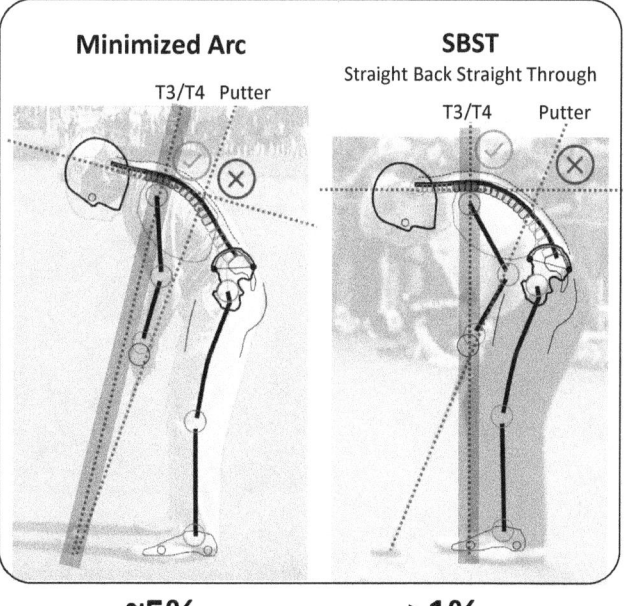

~100% of players using long putters are on a one plane posture

Even though the minimized arc and SBST postures both reduce the arc of the putting stroke, they are not very common on tour. The nearly mythical Straight Back Straight Through stroke (after the 2000 Pelz book) became the goal of many amateurs, but the two-plane neutral arc posture dominates the tour.

Both the one-plane (long putter) and two-plane (short putter) have the T3/T4 axis of rotation aligned with the plane of rotation of the putter shaft (the two left figures above).

Our analysis will concentrate on the T3/T4 joint of the upper thoracic spine as the biomechanical axis of rotation in the modern putting stroke.

The <u>Two-Plane</u> Neutral Arc Putting Posture (used by most tour players today)
With T3/T4 axis/plane aligned with the putter shaft plane; with a near 70° standard lie angle putter.

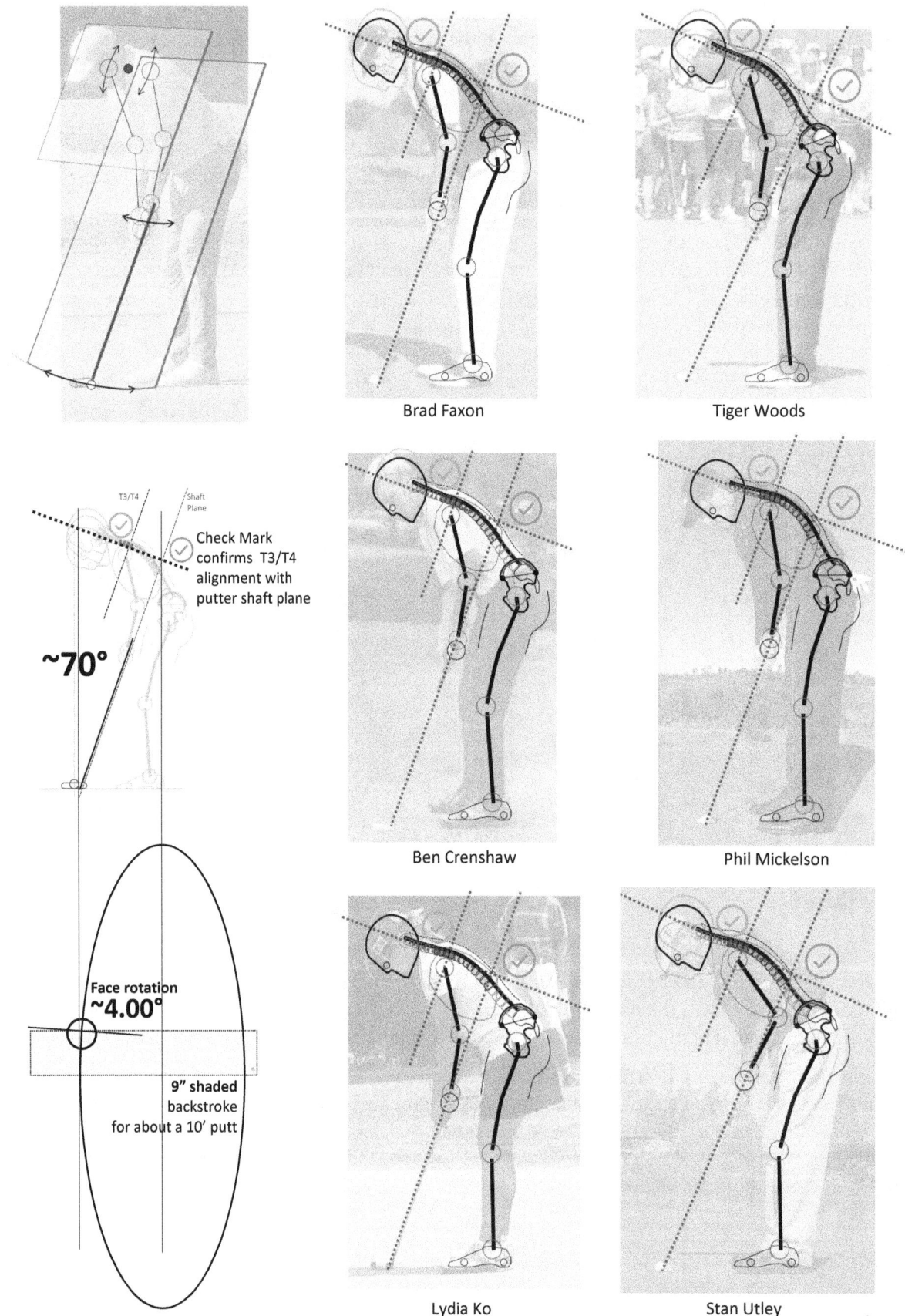

The <u>One-Plane</u> Putting Posture (used with long/broomstick putters)

With the T3/T4 axis/plane aligned with the putter head and ball; with a near 80° posture and an upright lie angle putter is used by a few tour players. This is **NOT** Straight Back Straight Through, the arc is **MINIMIZED** but not eliminated.

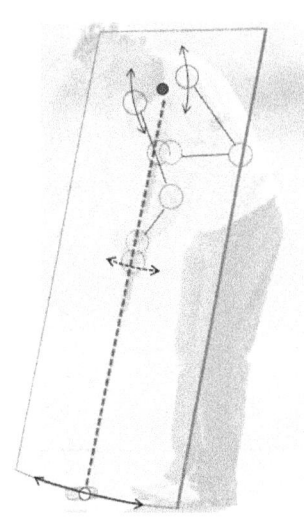
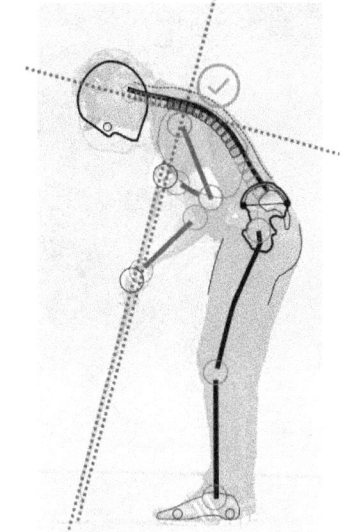
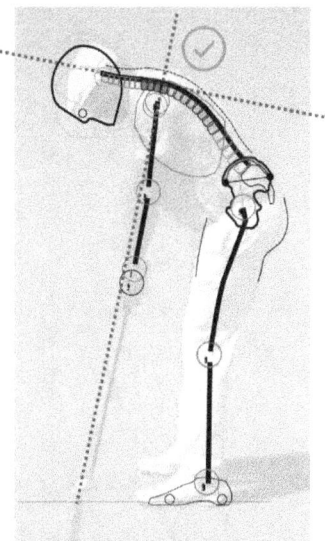

Bernhard Langer   Bryson DeChambeau

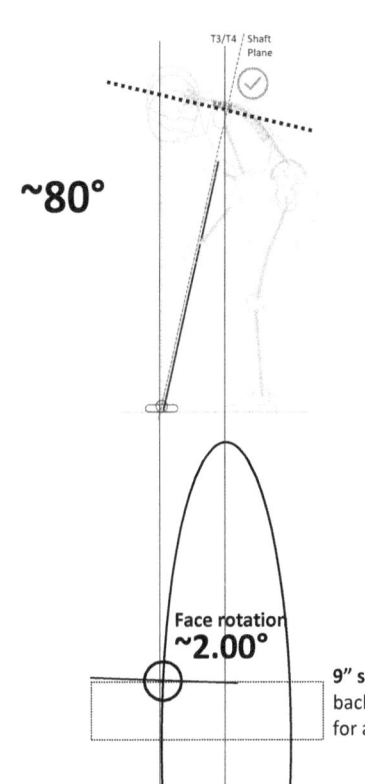
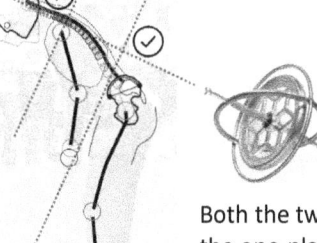

Both the two-plane posture (previous page) and the one-plane plane posture (this page) have the thoracic axis of rotation aligned with the axis of rotation of the putter. This alignment is STABLE and gyroscopic.

This STABILITY is the advantage of the long putter. The anchoring controversy has obscured the underlying posture advantage

An advantage of the one-plane long putter posture is reducing the arc, anchoring is not required. Scott McCarron says that he would not go back to anchoring even if it were allowed again in the future.

An additional advantage of the one-plane long putter posture is freedom of rotation on the T3/T4 Thoracic axis with the putter rotating on its plane of rotation.

The inherent STABILITY of the two-plane and one-plane postures is the reason for their dominance on tour

The Minimized Arc Putting Posture (used by some tour players)

With the T3/T4 axis/plane aligned with the putter head and ball; with a near 80° posture and an upright lie angle putter is used by a few tour players. This is **NOT** Straight Back Straight Through, the arc is **MINIMIZED** but not eliminated.

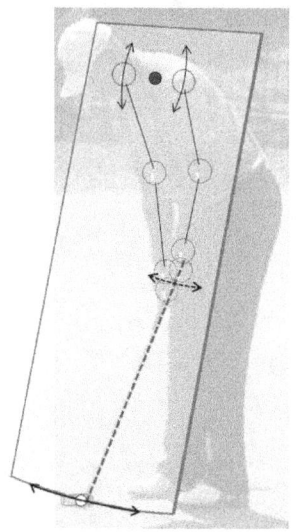

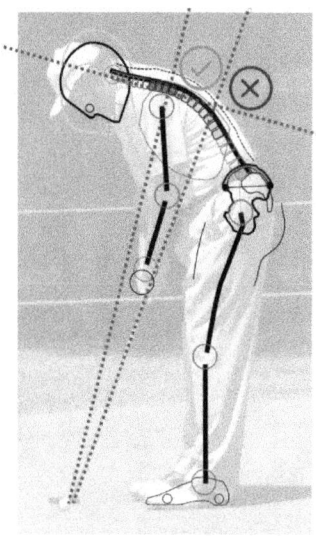
Loren Roberts

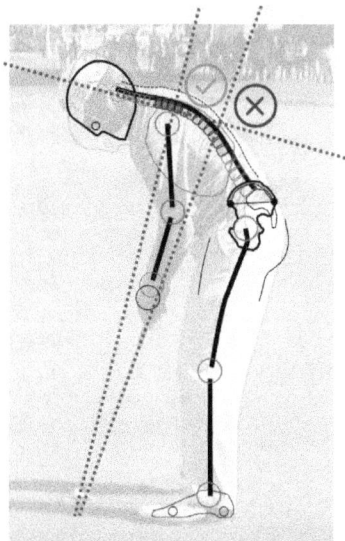
Steve Stricker

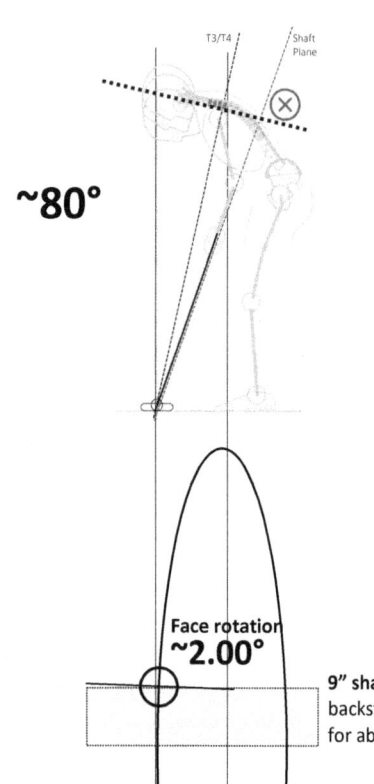

The minimized arc posture (this page) and the SBST posture (next page) both have the advantage of reducing the arc of the putter, reducing face angle variation.

But they are not widely used on tour. The two-plane posture dominates the tour, with over 95% of tour players.

The obvious advantage of reduced arc from the minimized arc and SBST postures is not enough to cause tour players to adopt them widely. The Dave Pelz excitement about the SBST stroke in 2000 produced a nearly mythological bias toward the SBST stroke as a goal.

The Straight Back Straight Through (SBST) Posture

With the T3/T4 axis/plane vertical. Very few players have actually used the Straight Back Straight Through putting posture and stroke recommended by Dave Pelz. Jack Nicklaus did. There are very few others; and none as successful as Jack Nicklaus.

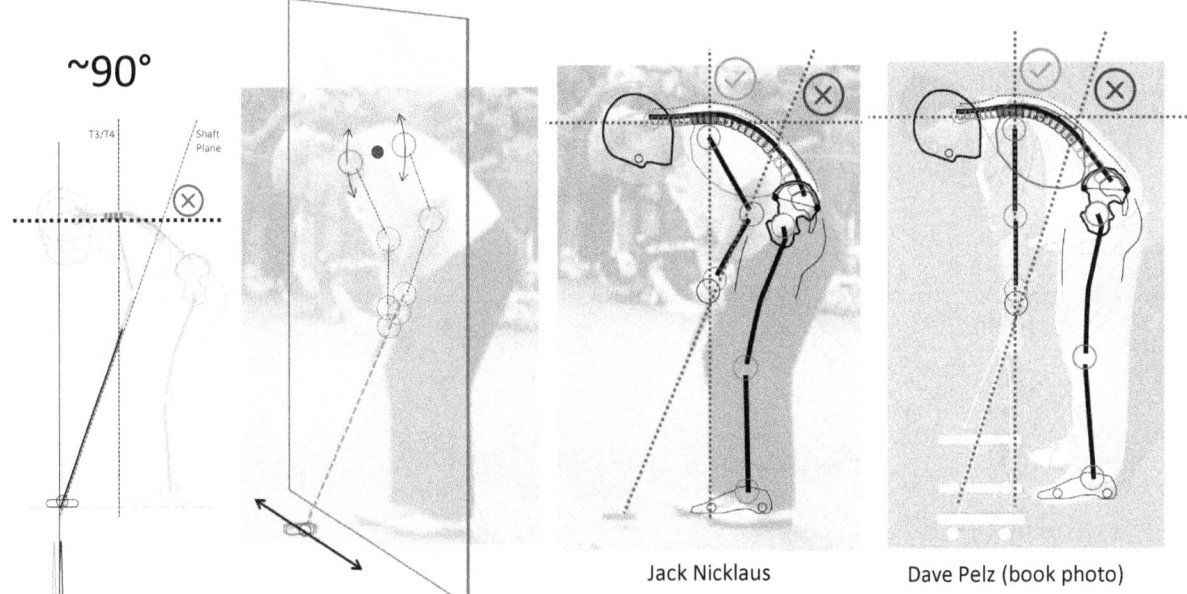

Jack Nicklaus   Dave Pelz (book photo)

Jack wandered back and forth over the SB/ST posture/path; sometimes the black path/posture; sometimes the gray.

Face rotation ~0.00°

9" shaded backstroke for about a 10' putt

The right path here is the open-to-closed path from standing just a little taller. The left path here is the closed-to-open path from bending over just a little more.

Both would look like, and are indeed very close to, straight back straight through.

Jack Nicklaus was very close to the "Straight Back Straight Through" putting stroke.

His distinctive crouched over posture had his T3/T4 axis very close to horizontal. He would some days stand a little taller, and other days bend over a little more.

Jack called this "square-to-square". In his famous book Golf My Way, he said "Today I am sometimes a square-to-square putter, I am sometimes an open-to-closed putter, I am sometimes a closed-to-open putter. And I couldn't tell you which it will be from week to week."

It is clear from this that he stayed close to this 90° posture and very little face rotation.

Dave Pelz was very influential in establishing the straight back straight through stroke as a goal; and clearly understood that it required a posture like Jack Nicklaus'

The "tabletop" posture paid homage to Jack Nicklaus but was not successful.

The problems are clear from a look at the conflicting axes of rotation and planes of rotation.

Wie needed a lot of hands/wrists/forearms input to keep the putter on her intended line.

The tendency was likely a strong closed-to-open stroke due to the tilt of the T3/T4 axis.

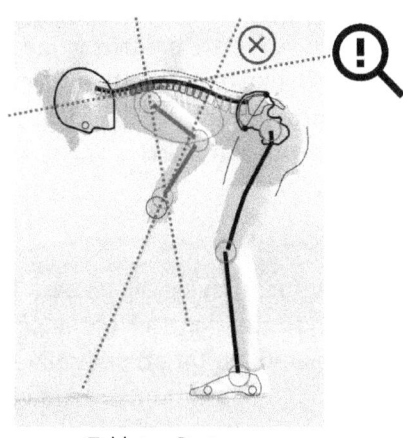

Tabletop Posture

The Strong Arc of the old fashioned "wrist cock/release" stroke

The figures below demonstrate the dramatic increase in the arc of the putting stroke when the hands/wrists/forearms are active in the stroke.

Wrist Cock/Release Plane and Arc

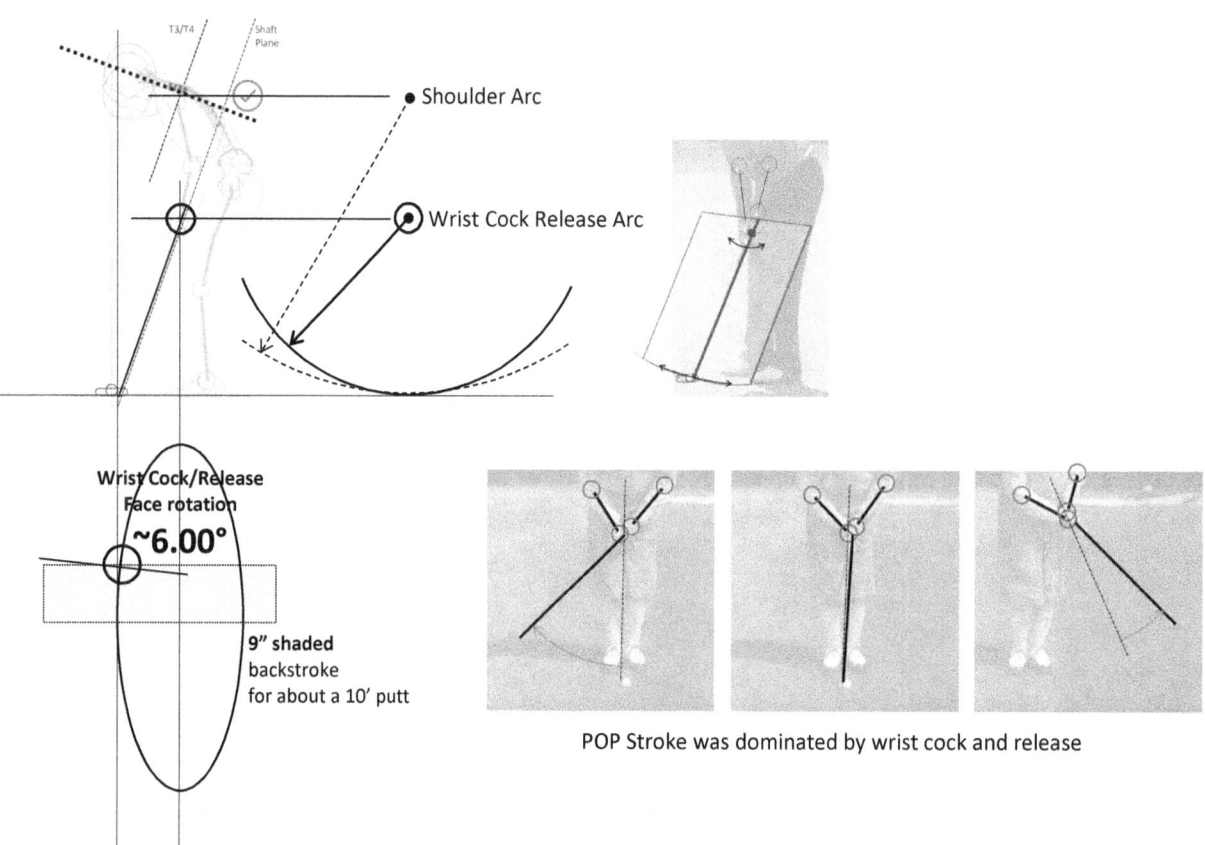

POP Stroke was dominated by wrist cock and release

The old-fashioned wrist stroke developed in a time with shaggy very slow greens (by today's standards) and was used with a very lofted putter (by today's standards). 200 years ago a players might carry 3 or 4 putters of varying loft for putts (actually more like chip and run shots) from longer distances. The fast greens we all play on today have allowed the development of the shoulder turn only modern putting stroke with much lower lofted putters.

Players of all skill levels fight to reduce hand/wrist/forearm actions in their putting today. Hand/wrist/forearm action can add 6° or more of face rotation (over the neutral arc face rotation). When working with a player with excessive face rotation, putter fitters have for many years "calmed the strokes rotation" by adding toe hang, which increases the putter's resistance to twisting (rotating on the axis of the shaft).

NOTE:
Before moving on, go back and look carefully at the tilt of the plane of rotation and the face rotation indicated for each of the postures (at the left side of each of previous pages). Note that the tilt of the plane of rotation determines the amount of face rotation in the stroke. A more detailed look at the plane of rotation, the circular path of the putter and its elliptical arc follows.

Biomechanical Putting Posture
5.2 The Elliptical Arc of the Putting Stroke

Before we look at the biomechanical issues related to freedom of rotation, we are going to look at the geometry of the arc of the putting stroke and its geometric relationship to posture.

The Elliptical Arc (on the ground) is determined by the tilt of the plane of rotation
The circular path of the putter head on the plane of rotation is the reason that the path of the putter follows on the ground is elliptical. The shadow of a tilted circle on the ground when the sun is straight overhead is an ellipse. The putter remains square to the arc, but traces a curving path.

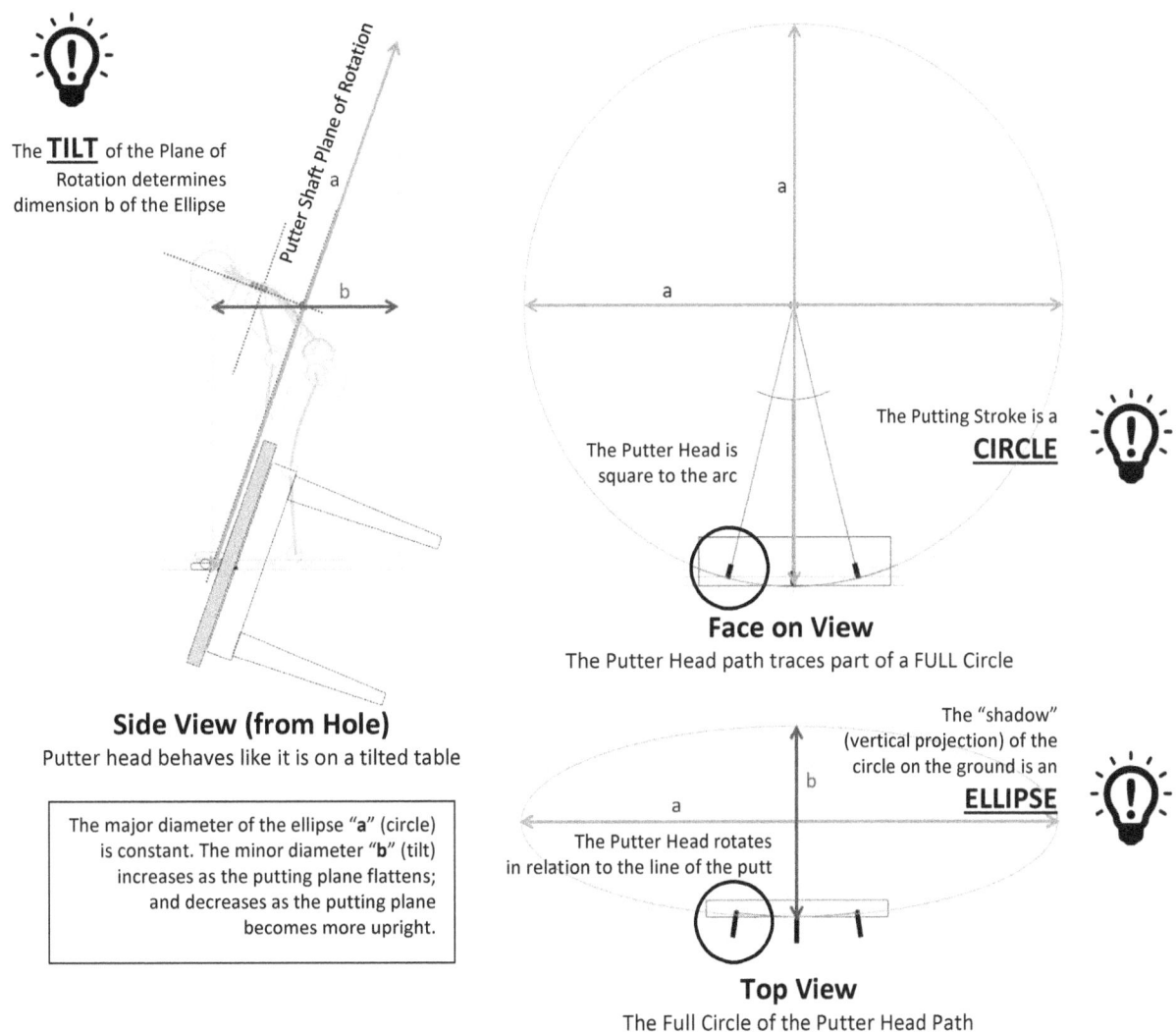

The neutral rotation of the putter head (no hands/wrists/forearms) is square to the arc. The elliptical arc causes the putter face to rotate in relation to the line of the putt. The amount of face rotation (in relation to the line of the putt) depends on the tilt of the plane of rotation (posture and shaft lie angle).

The Plane of the Putting Stroke Determines the Arc of the Stroke

and the amount of putter face rotation. Shown below is face rotation at 9" backstroke.

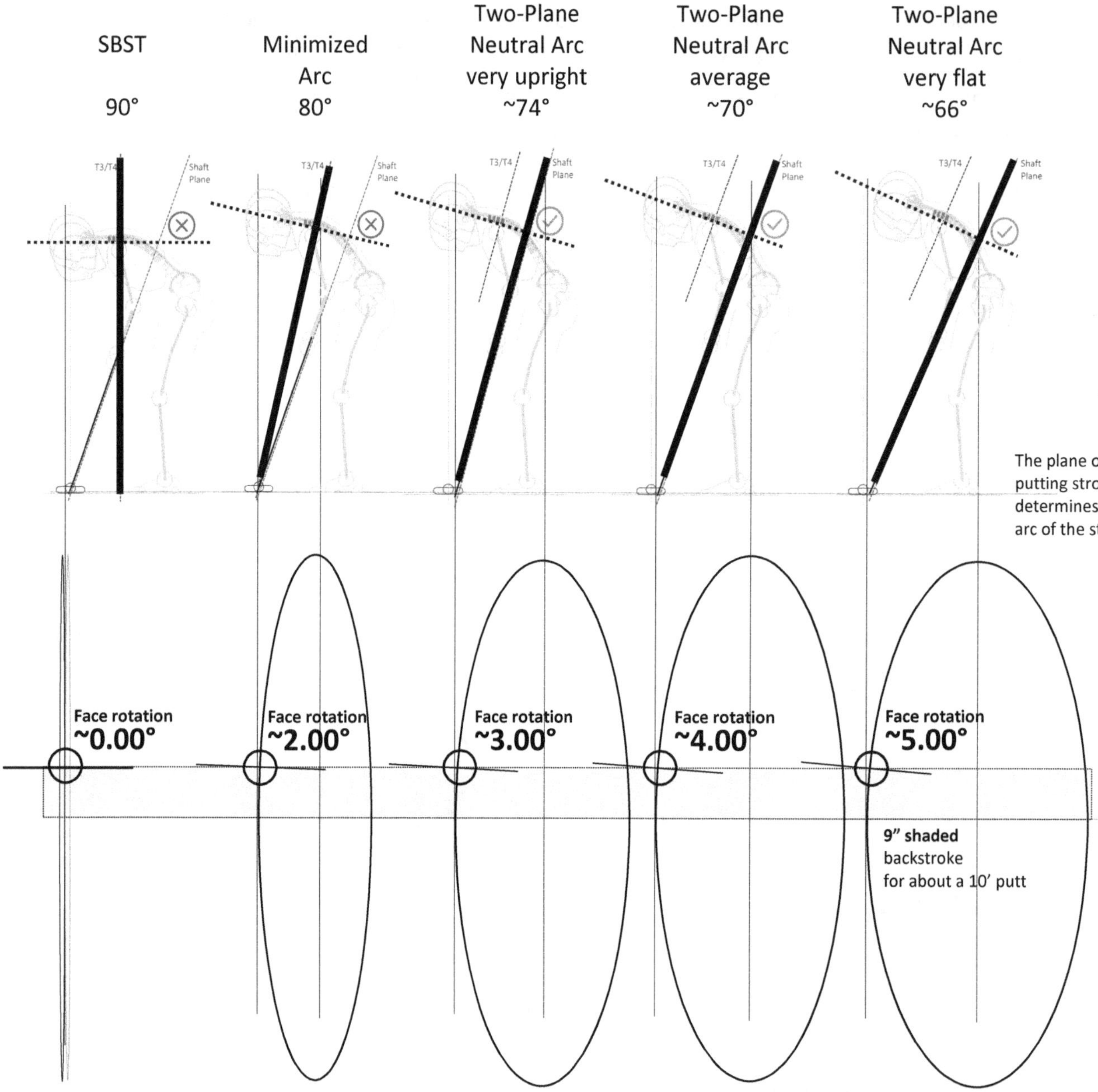

SBST	Minimized Arc	Two-Plane Neutral Arc very upright	Two-Plane Neutral Arc average	Two-Plane Neutral Arc very flat
90°	80°	~74°	~70°	~66°
Face rotation ~0.00°	Face rotation ~2.00°	Face rotation ~3.00°	Face rotation ~4.00°	Face rotation ~5.00°

The plane of the putting stroke determines the arc of the stroke

9" shaded backstroke for about a 10' putt

← Upright plane = **LESS** face rotation

Flatter plane = **MORE** face rotation →

An Important Question:

Dave Pelz's Putting Bible (2000) recommended the zero arc SBST putting posture and stroke. With the clear advantage of less face rotation with the zero arc stroke, why have so few tour players chosen to use it? Over 95% of tour players chosen the neutral arc posture, why?

The answer is in the biomechanics of the spine and physics of the pendulum and gyroscope.

Arc Varies with Lie Angle and Player's Height

Arc Varies with Lie Angle

The wide range of biometrically sound putting postures is shown below. Correct Alignment is possible with all of the postures below, when the posture and putter lie angle are aligned; and the putter is the proper length.

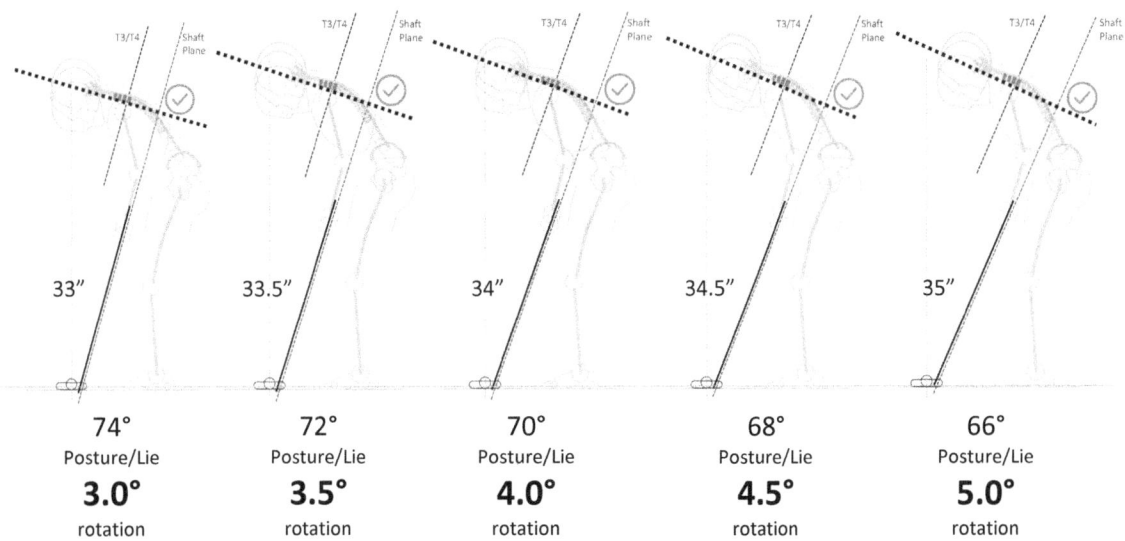

33"	33.5"	34"	34.5"	35"
74° Posture/Lie	72° Posture/Lie	70° Posture/Lie	68° Posture/Lie	66° Posture/Lie
3.0° rotation	3.5° rotation	4.0° rotation	4.5° rotation	5.0° rotation

Lie angle and Putter Length are directly determined by posture. Putter lengths shown above are imprecise and provided only as generalized relative values.

Generally, our experience has been that players have longer putters than they need. Be careful with the length of your putter. Let your arms hang freely with no tension.

Arc Varies with the Player's Height

In the examples below height varies but the 70° standard putter lie angle (and posture) kept constant. All other things being equal, a taller player will have less arc; and a shorter player will have more arc.

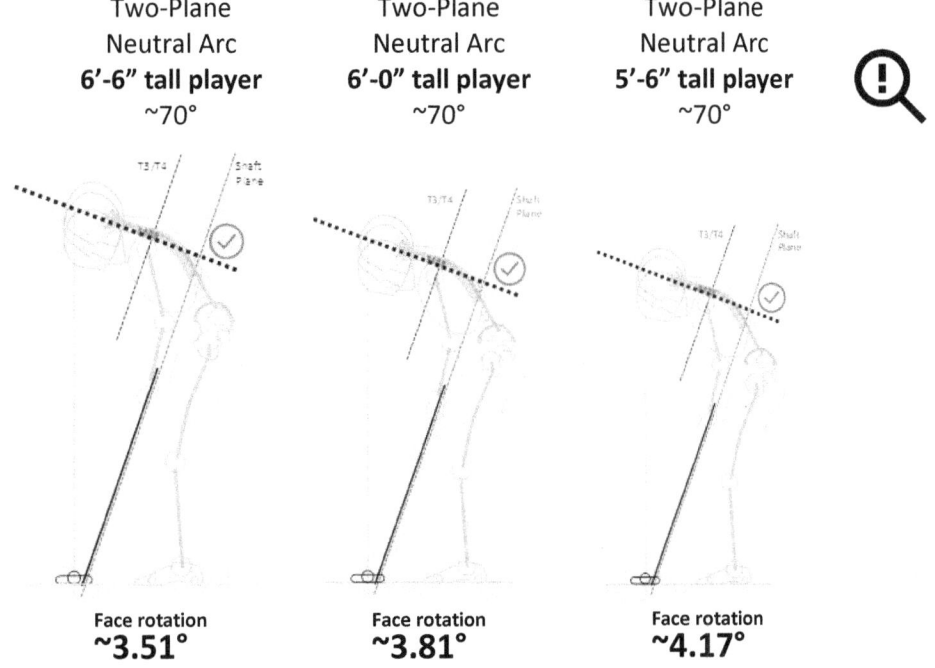

Two-Plane Neutral Arc 6'-6" tall player ~70°	Two-Plane Neutral Arc 6'-0" tall player ~70°	Two-Plane Neutral Arc 5'-6" tall player ~70°
Face rotation ~3.51°	Face rotation ~3.81°	Face rotation ~4.17°

Biomechanical Putting Posture

5.3 Square to the Arc Face Rotation Varies with Posture

Dave Pelz recommended the Straight Back Straight Through (SBST) putting stroke in his Putting Bible published in 2000. The nearly mythical SBST stroke is easy to visualize.
Square means "aligned at 90° to".

In the SBST stroke,
the putter is Square to the Line of the Putt.
The SBST stroke has no arc

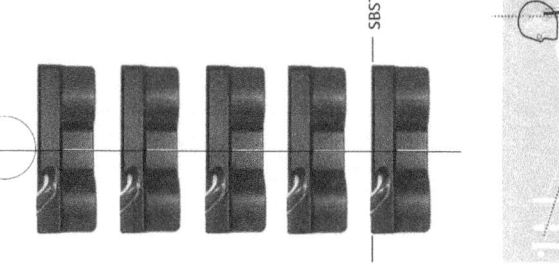

The SBST stroke is much more talked about than used. Dave Pelz was very careful to explain the posture required to produce a SBST stroke (above right). But the ideal of the SBST through took on a life of its own, divorced from the required posture. Note that less than 1% of tour players are in a posture like this. Nearly ALL Tour players are in a putting posture that produces an arc.

We are introducing the concept of the neutral arc, the arc naturally produced by the tilt of the putting stroke plane in the chosen putting posture. Almost everyone has an arc putting stroke because of their putting posture and the tilted plane of their putting stroke. The example illustrated below shows a 9" backstroke with 4° of rotation. This is the approximate neutral arc rotation for an average tour player. The putter stays Square to the Arc.

In the Two-Plane Neutral Arc
and all postures producing an arc
the putter is Square to the Arc.

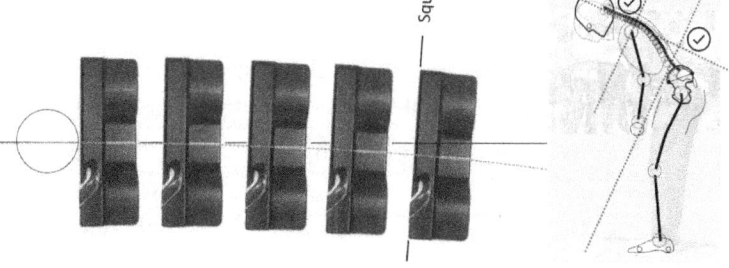

The concept of Square to the Arc is important to understand. The average tour player putting stroke looks like the illustration immediately above. Launch monitor data indicates that tour players allow this square to the arc rotation and do not attempt to manipulate the putter to a SBST path. But many amateurs are still trying to create a SBST putter path from a posture that naturally produces an arc or adding putter rotation with unwanted hand/wrist action.

NOTE
The square to the neutral arc rotation for a 9" backstroke varies with the putting postures below:

Biomechanical Putting Posture
5.4 Rotational Stability in the Putting Stroke

The rotational stability of the putter on the plane of the putter shaft
provides <u>rotational stability like a gyroscope</u>.

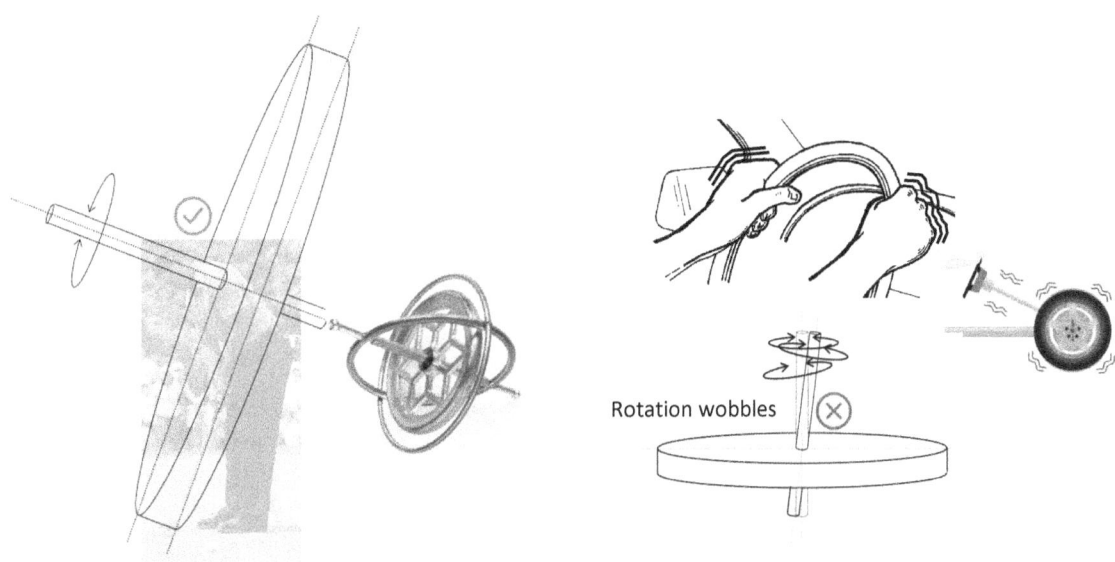

A gyroscope spins smoothly on its axis of rotation and even "defies gravity" because of the special stability of rotational inertia. If the axis of rotation is not perpendicular, or not centered, the rotation would wobble, like an unbalanced wheel on your car. The more off perpendicular, or off center, the axis is, the more severe the wobble will be. Wobble in a putting stroke means the hands will not be quiet; the putter shaft will try to stay on plane, but cannot. The degree to which it is pulled off the plane will be felt in the hands; like a shaking steering wheel in a car with a badly aligned wheel (but much more subtle).

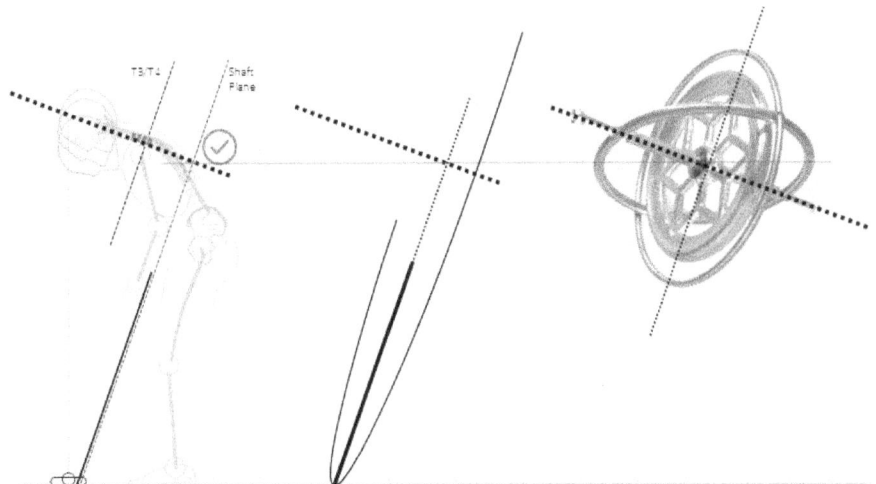

The alignment of the T3/T4 axis/plane with the plane of putter shaft
<u>provides the least possible resistance to free rotation of the modern putting stroke around the spine</u>. This freedom of rotation creates less resistance to a pendulum tempo with constant acceleration. The neutral arc posture with the alignment of the spinal axis of rotation and the plane of rotation of the putter shaft is therefore likely to be the posture that most easily produces the pendulum tempo in the previous chapter.

The T3/T4 Axis of Rotation and Plane of the Putting Stroke

The Putting Plane is different from the full swing plane. Both the head and hips are fixed in the putting stroke; only the head is fixed in the full swing. The axis and center of rotation of the putting stroke is not clear at a glance because the gesture of the back as a whole is so strong; but the axis of rotation of the putting stroke is in the upper back between the shoulders approximately at the joint between vertebrae T3 and T4. The axis of rotation in the full swing is much lower in the back around T7/T8

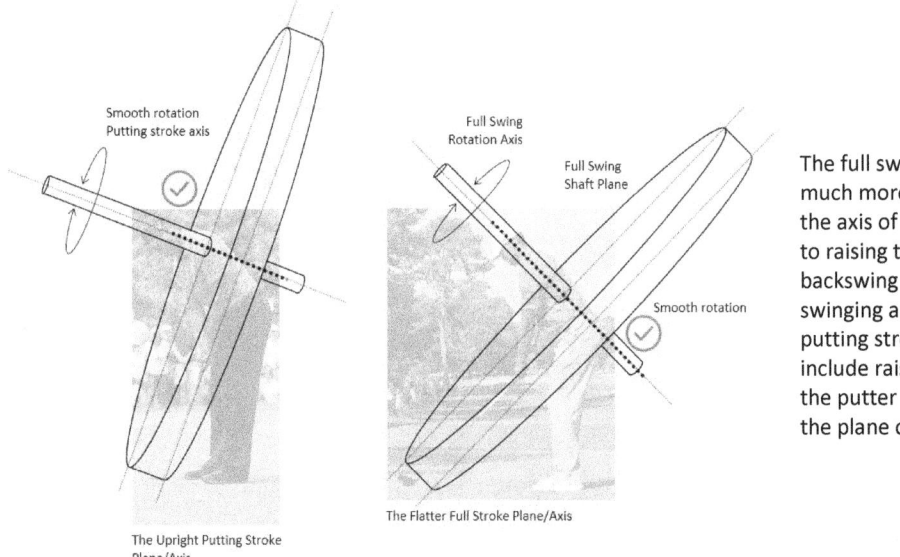

The full swing plane is much more vertical than the axis of the swing due to raising the arms in the backswing (like when swinging an axe); the putting stroke does not include raising the arms so the putter shaft stays on the plane of the axis

To allow the putter to swing smoothly (like a gyroscope on its axis of rotation); and for the shoulders to rotate smoothly on their axis of rotation; the putting plane and the spinal axis of the putting stroke rotation must be aligned; like the center figure below. Shoulder rotation is the steering wheel of the putting stroke.

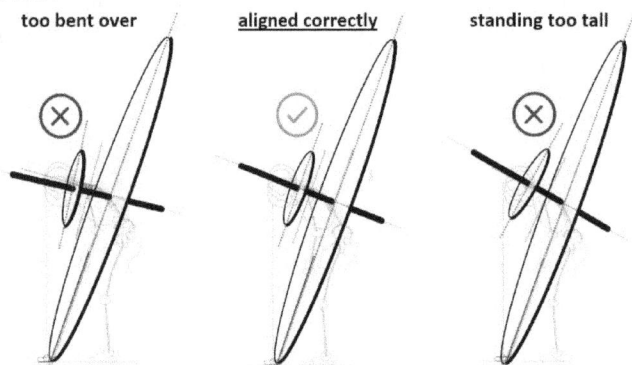

Only in the center figure here can the putter be swung on the plane of the putter shaft without scapular manipulation.

In the left figure the putter will be above the plane and tend to take a closed-to-open path. In the right figure the putter will be below the plane and tend to take an open-to-closed path.

The small errors in alignment in the right and left figures above cause the putter shaft to rotate off its plane or create a need for non-rotational shoulder movement which is very difficult to control.

The putter will rotate most smoothly on the plane of its shaft, like a gyroscope. The spine can be aligned with this axis and plane of rotation.

Rocking the shoulders on the axis and plane of the T3/T4 joint with very limited hand/wrist/forearm action is the motion of the modern putting stroke. The T3/T4 axis allows adequate rotation for most putts. The faster greens today allow this modern stroke.

Biomechanical Putting Posture
5.5 Biomechanical Rotation on the Putting Stroke Axis

The biomechanics of the upper thoracic spine (T3/T4), the shoulder girdle and the connecting musculature is important to understand.

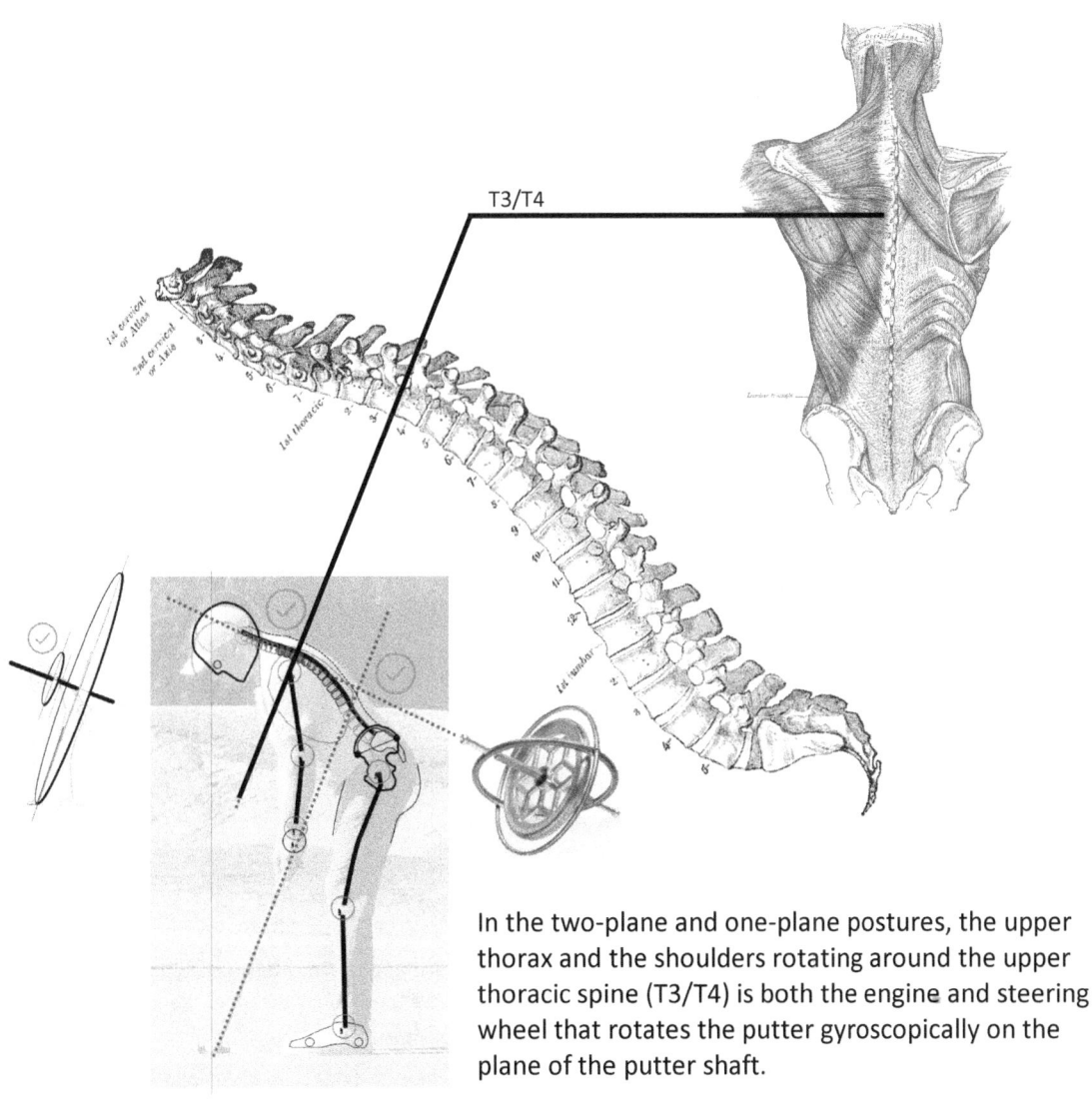

In the two-plane and one-plane postures, the upper thorax and the shoulders rotating around the upper thoracic spine (T3/T4) is both the engine and steering wheel that rotates the putter gyroscopically on the plane of the putter shaft.

The two planes of the Two-Plane Neutral Arc posture are:

Plane 1) the shoulders rotate around the T3/T4 axis and on its plane

Plane 2) the putter rotates (gyroscopically) on the plane of the putter shaft

The Biomechanical Confirmation for the T3/T4 Putting Stroke Axis

In a putting stroke, the hips and the lowest lumbar vertebrae and the head and the upper cervical vertebrae are fixed. The remainder of the lumbar and cervical vertebrae and all of the thoracic vertebrae rotate with variable ranges of axial rotation. The cervical vertebrae and upper thoracic vertebrae are extremely flexible and allow a lot of rotation. The lumbar and lower thoracic vertebrae are significantly less flexible and allow much less rotation.

The location of the shoulders as part of the biomechanical structure supporting a putting stroke makes the T3/T4 joint the natural center of the putting stroke. The vertebrae above and below T3/T4 have approximately the same total range of rotation (67° to 68°). The T3/T4 axis is biomechanically free.

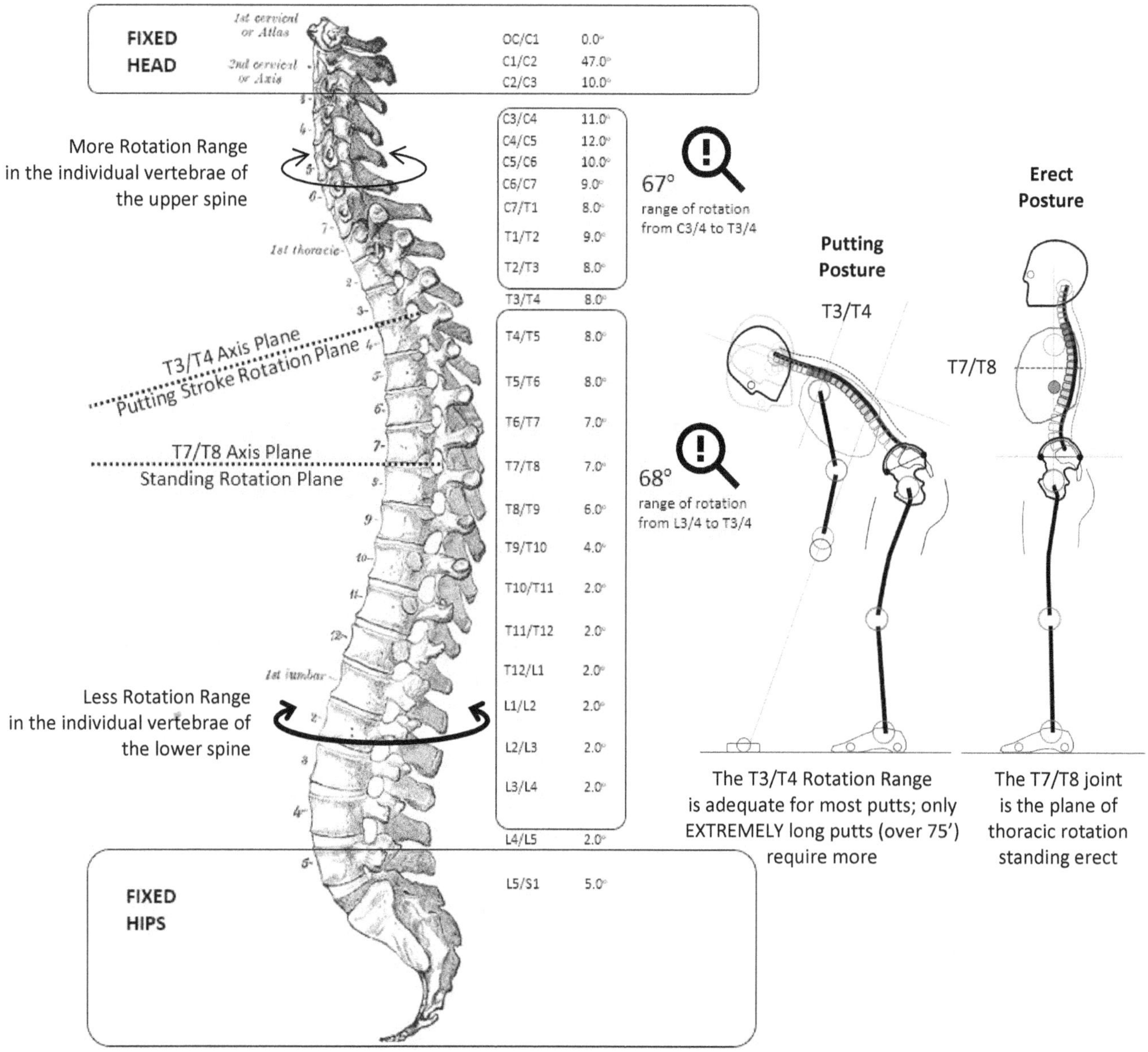

Range of Axial Rotation

Joint	Range
OC/C1	0.0°
C1/C2	47.0°
C2/C3	10.0°
C3/C4	11.0°
C4/C5	12.0°
C5/C6	10.0°
C6/C7	9.0°
C7/T1	8.0°
T1/T2	9.0°
T2/T3	8.0°
T3/T4	8.0°
T4/T5	8.0°
T5/T6	8.0°
T6/T7	7.0°
T7/T8	7.0°
T8/T9	6.0°
T9/T10	4.0°
T10/T11	2.0°
T11/T12	2.0°
T12/L1	2.0°
L1/L2	2.0°
L2/L3	2.0°
L3/L4	2.0°
L4/L5	2.0°
L5/S1	5.0°

FIXED HEAD

More Rotation Range in the individual vertebrae of the upper spine

T3/T4 Axis Plane
Putting Stroke Rotation Plane

T7/T8 Axis Plane
Standing Rotation Plane

Less Rotation Range in the individual vertebrae of the lower spine

FIXED HIPS

67° range of rotation from C3/4 to T3/4

68° range of rotation from L3/4 to T3/4

Putting Posture — T3/T4

The T3/T4 Rotation Range is adequate for most putts; only EXTREMELY long putts (over 75') require more

Erect Posture — T7/T8

The T7/T8 joint is the plane of thoracic rotation standing erect

Stand in the two postures shown at right above. First stand erect and rotate your thorax. You can feel the rotation centered at about T7/T8. Then take the putting posture and rotate at your shoulders. You can feel the rotation centered at T3/T4 at the shoulders.

Shoulder (Scapular) Movement Required if T3/T4 is NOT on Plane

A player in the "Extreme Upright Posture" or in "Minimized Arc Posture" cannot rotate on the putter shaft plane without shoulder (scapular) movement; which is very hard to control. Moving the shoulders on the putter shaft plane from these postures is not as simple as it looks.

The path of movement in the complex shoulder girdle causes the shoulder to move asymmetrically; and both Elevation and Depression, and Protraction and Retraction are required to accomplish the movements. These together move the shoulders off of the line of the putting stroke and also move the center line of the shoulders away from the center of the back.

These movements must be balanced very precisely to cause no change to the line of the stroke!
AND
The freedom of rotation in the neutral arc posture will more naturally and freely allow the development of your personal pendulum tempo.

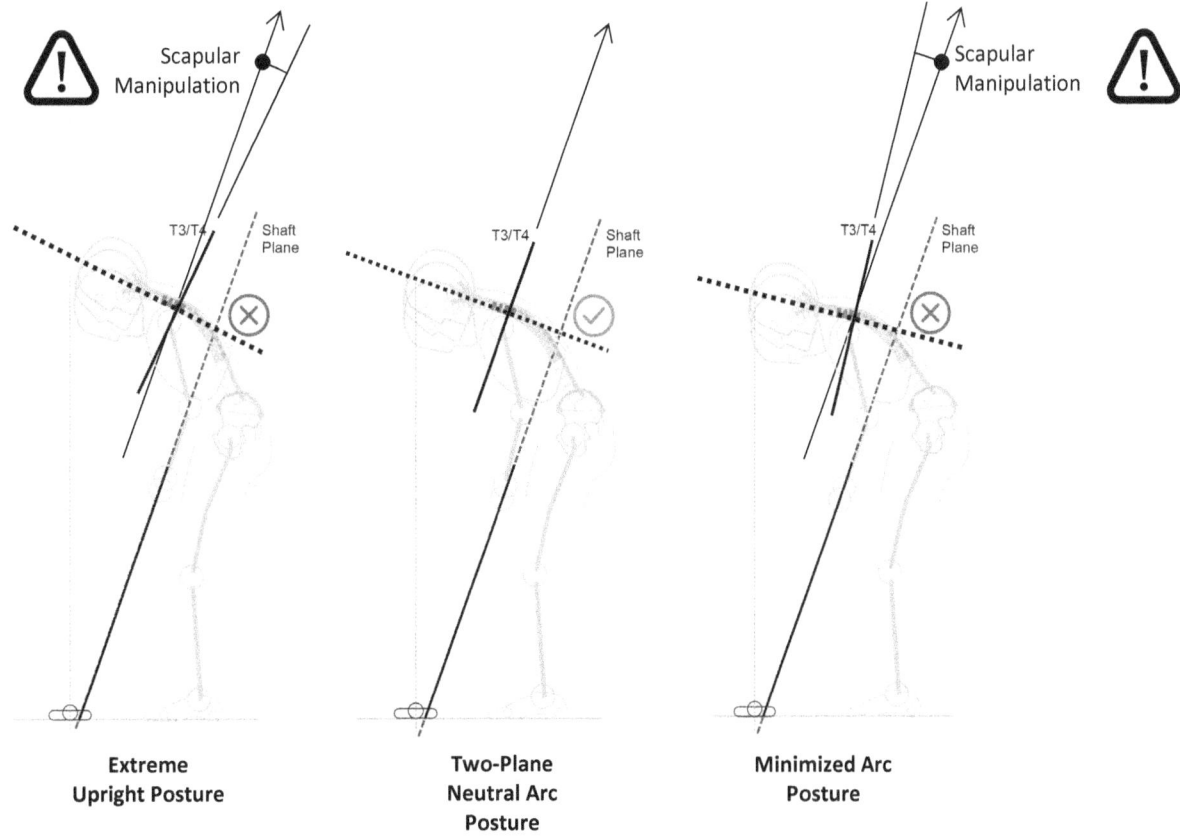

Extreme Upright Posture

Two-Plane Neutral Arc Posture

Minimized Arc Posture

NOTE:
The natural selection of the two-plane neutral arc posture by most tour players is likely related to the ease of developing the pendulum putting tempo in this posture. Some tour players use the one-plane minimized arc posture. The reduced arc advantages of the minimized arc are likely overwhelmed by relative difficulty of developing the pendulum putting tempo in this less efficient posture. Tour players have overwhelmingly chosen the two-plane neutral arc posture. The extreme upright posture has been rejected by ALL tour players.

The analysis below provides an explanation of why the players on the PGA Tour have overwhelmingly chosen the two-plane neutral arc posture. They have made this "choice" on the basis of feel and results. The analysis below provides the science behind their choice.

The Movement of the Shoulder Girdle
The shoulder joint is remarkably mobile, and floating in muscular balance

The shoulder is the most mobile joint in the human body. The pectoral girdle, the Clavicle and Scapula, is attached to the frame of the body at the Sternum by the sternoclavicular joint (a true joint); otherwise the shoulder floats relatively free. The scapula, the shoulder blade, rides on the back of the rib cage. The shape of the rib cage (below center) is a surprise to many. It is the free floating, very mobile shoulders, that give the torso the familiar wide at the shoulders narrow at the hips look.

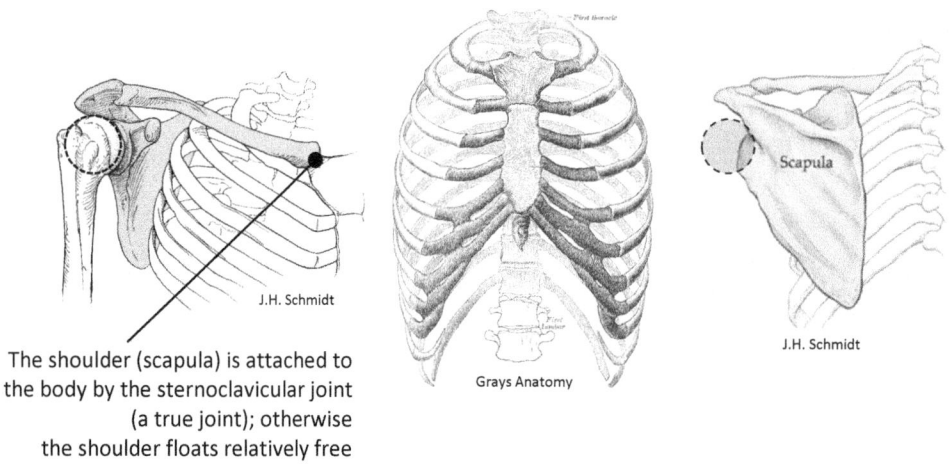

The shoulder (scapula) is attached to the body by the sternoclavicular joint (a true joint); otherwise the shoulder floats relatively free

The muscles of shoulder movement

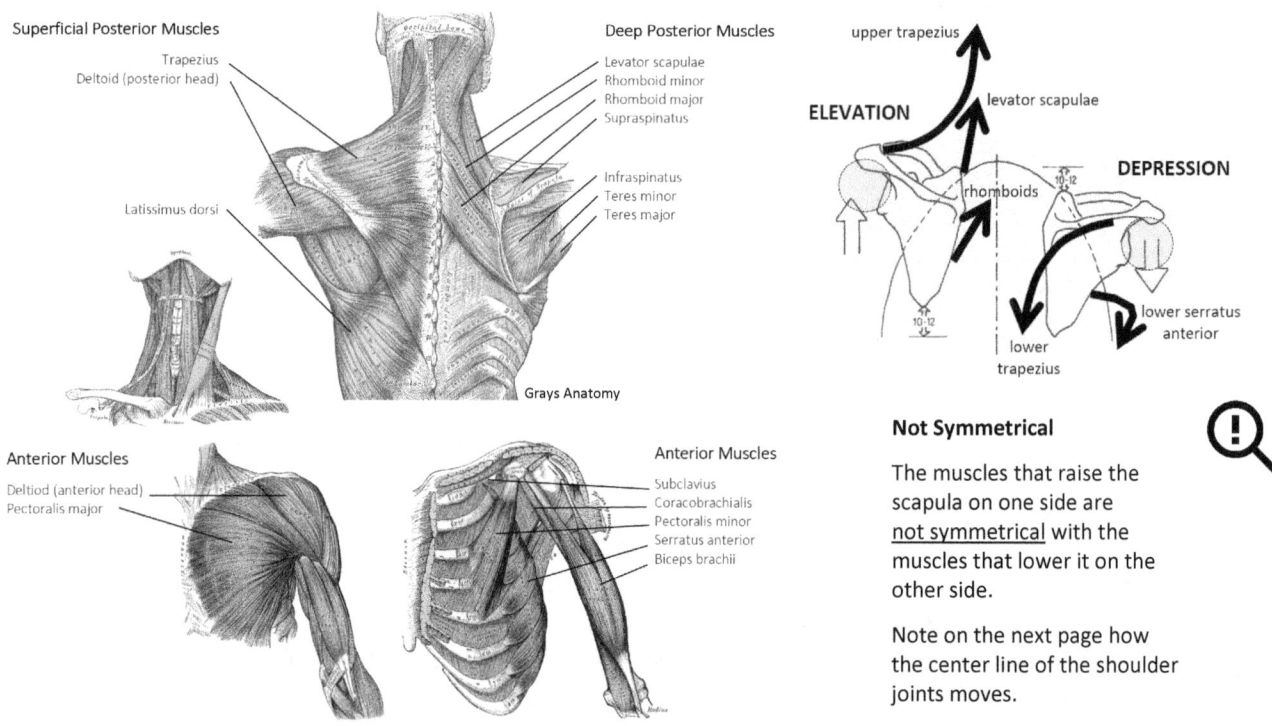

Not Symmetrical

The muscles that raise the scapula on one side are not symmetrical with the muscles that lower it on the other side.

Note on the next page how the center line of the shoulder joints moves.

The musculature of the shoulder is very complex and very difficult to control in a putting stroke.

Shoulder Movement & Musculature is Asymmetrical and Difficult to Control

Not only are the muscles asymmetrical and difficult to control, but the path of the shoulder around the back of the rib cage is curved in three dimensions, and asymmetrical.

NOTE:
The center line between the shoulders shifts from the center of the back, when you utilize shoulder manipulation

Elevation and Depression

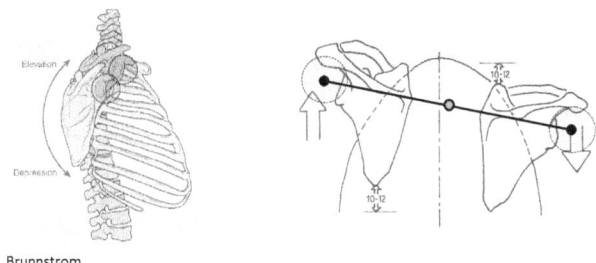

Brunnstrom

Elevation and depression movements follow the curve of the rib cage vertically and are parabolic laterally on the surface of the back.

Note that the center line between the shoulders has shifted from the center of the back!

Protraction and Retraction

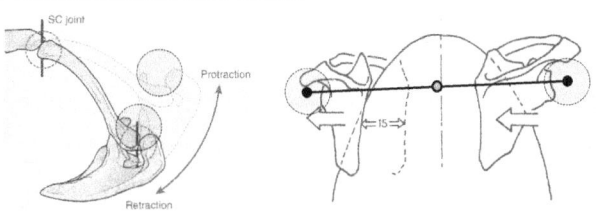

Brunnstrom

Protraction and retraction follow the curve of the rib cage horizontally but depart from the parabolic elevation/depression curve on the surface of the back.

The geometry of these movements is very complex, and controlled by a large number of interacting muscles.

Rotation (with raising of the arm)

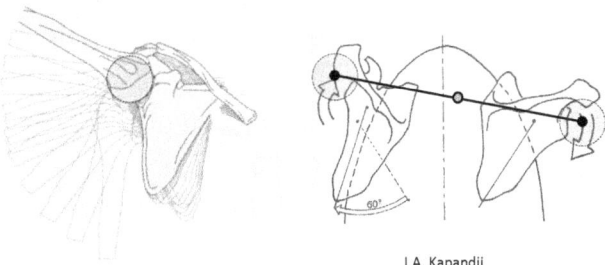

I.A. Kapandji

The scapula does not rotate with the first 30 degrees of arm movement, but then rotates quite a lot, putting the shoulder joint up adjacent to the ear.

The paths of the shoulders are curving in three dimensions and the

CENTER LINE BETWEEN THE SHOULDERS HAS SHIFTED FROM THE CENTER OF THE BACK!!!

The elevation, depression, protraction, retraction movements are anchored by the clavicle creating the parabolic arc; and shifting of the shoulder center line from the center of the back. This muscular manipulation is very difficult to control, and results in very erratic putting without extremely fine control; and lots of practice.

IT IS FAR EASIER TO LEAVE THE SHOULDERS IN A FIXED POSITION AND ROTATE AROUND THE SPINE!

Biomechanical Putting Posture

5.6 Biomechanical Fitting Protocol

Determine the Putting Posture you want to take

The two-plane neutral arc posture is by far the most common posture on tour. It is the most likely choice. But it is not the only choice. The minimized arc and zero arc SBST postures are reasonable to explore but require much more practice to maintain a free pendulum tempo.

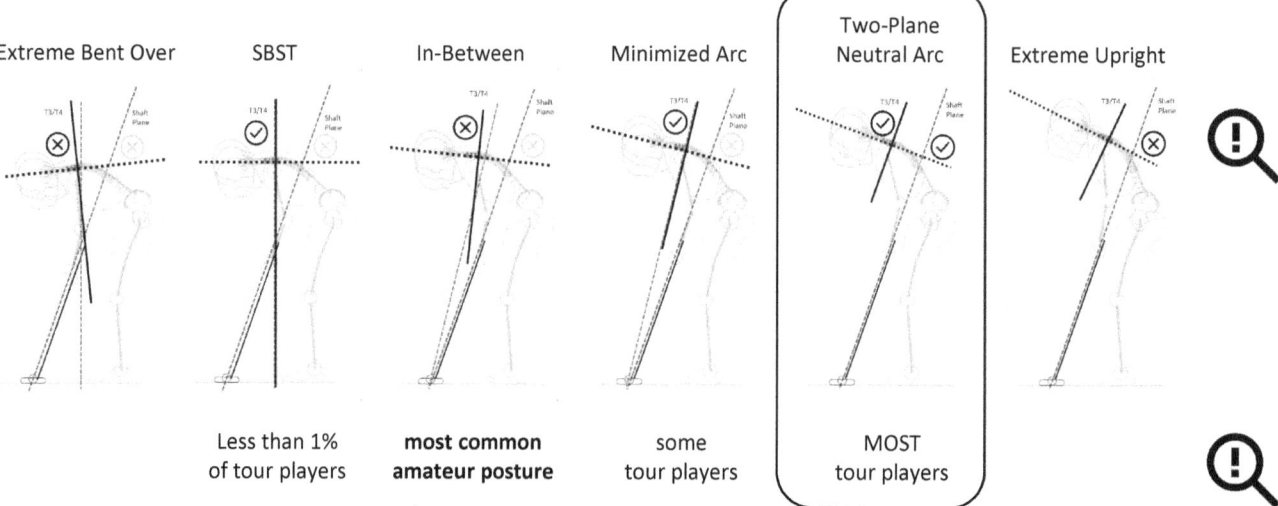

Tour players are almost all standing far taller than most amateurs. Tour players are almost all in the neutral arc posture. Even though the SBST posture is almost non-existent on tour, teachers and pundits in the golf industry today still claim that many players have a straight back straight through stroke. Players in the One-Plane "Minimized Arc" posture have less arc in their putting stroke and are often called straight back straight through putters. Confusing the issue further is the fact that many players add arc to their stroke with added hands/wrists/forearm rotational actions that create a much stronger arc. The widespread use of imprecise terminology has confused the discussion.

Notice putting posture every time you watch tour players on TV. Study the range of postures on the next few pages.

The Two-Plane Neutral Arc Posture is the overwhelming choice of tour players. There are not as many tour players in the minimized arc posture but they tend to be very good putters. Jack Nicklaus used the zero arc posture with GOAT success. His success alone should cause more exploration of this posture.

The extreme upright posture should be avoided. If you are in this posture you should move to the two-plane neutral arc posture.

The key to biomechanical fitting is discerning where the T3/T4 axis and rotation plane is oriented. Look back at the range of putting postures and the pictures of tour players at the beginning of this section.

Biomechanical Putting Posture

The Two-Plane Neutral Arc Posture uses a <u>Relaxed</u> "Athletic" Putting Stance

1

Start from an erect standing posture;

2

Bend over into the "ready" athletic stance. This is the stance of a baseball infielder ready to move in any direction toward the ball when it is hit. Strongly grounded. Center of gravity over the balls of the feet.

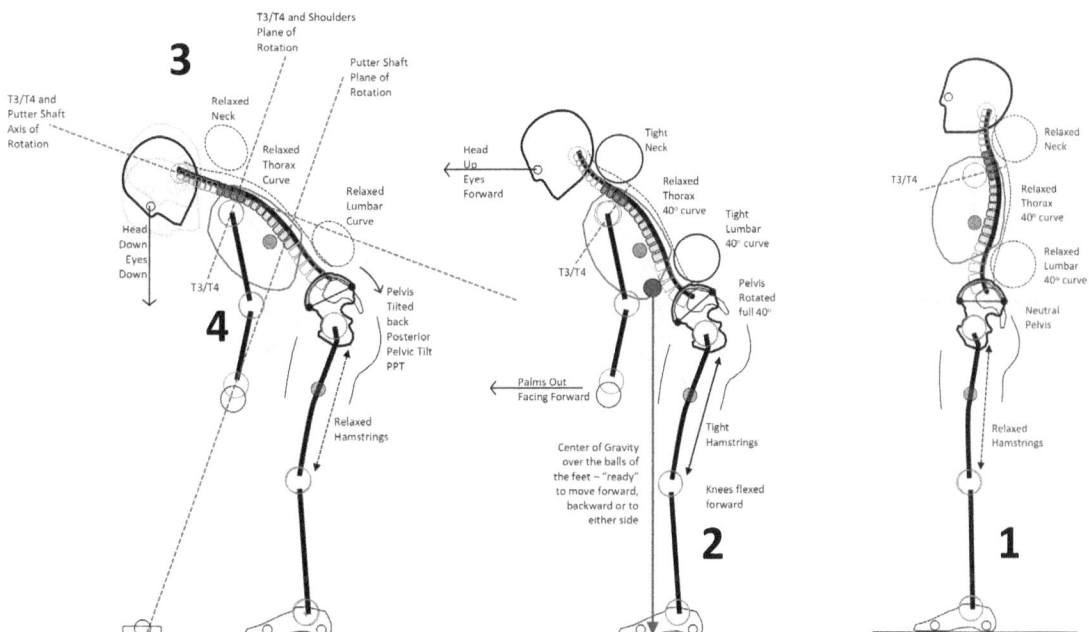

3

To evolve from this general athletic stance to the <u>relaxed</u> athletic putting stance you must relax your lower back a little. This lower back relaxation requires a subtle posterior pelvic tilt (lowering your butt a little). You also need to bend your head down to look at the ball. Relaxing your lower back and changing your pelvic tilt allows you to comfortably "bend over" into the <u>relaxed</u> athletic posture with your upper thoracic spine (the T3/T4 axis/plane) aligned with the plane of the putter shaft.

The relaxed feeling in the lower back comes from pelvic tilt. To straighten up you have to change your pelvic tilt. To relax you need to change your pelvic tilt. Most people have never thought about this, we do it naturally without thinking. Pelvic tilt controls the lumbar curve in the spine.

4

Finally, you need to let your arms hang freely with no tension, to establish the right putter length. If you have pulled your arms up towards your chest, bending your elbows more than in the relaxed position, you will feel you need a longer putter than is right for your posture.

Fine Tuning your Biomechanical Posture

Whatever posture you choose, you must carefully align your T3/T4 axis, adjusting from too upright (left below) or too hunched over (right below), to correctly aligned (center below).

Two-Plane Neutral Arc Posture Adjustment

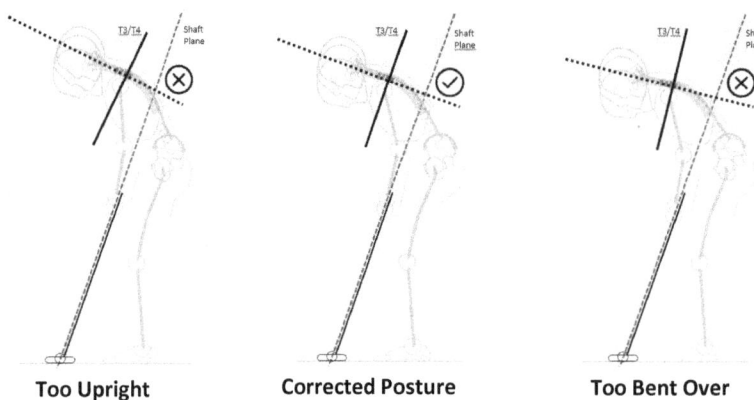

Too Upright   Corrected Posture   Too Bent Over

Minimized Arc Posture Adjustment

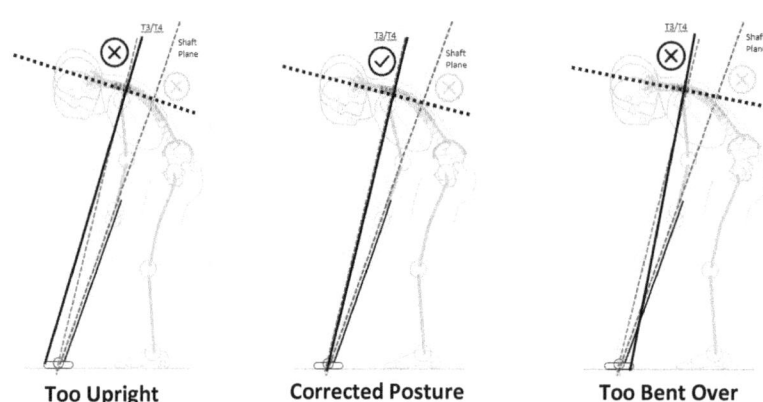

Too Upright   Corrected Posture   Too Bent Over

Zero Arc SBST Posture Adjustment

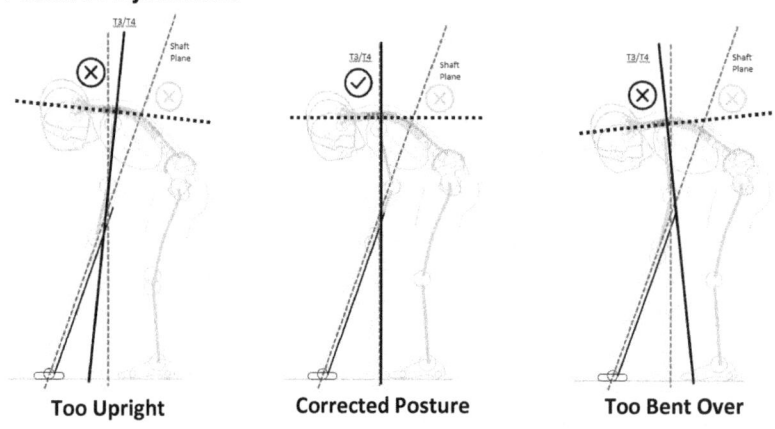

Too Upright   Corrected Posture   Too Bent Over

You can make these adjustments in a mirror or working with your teacher or with a friend. You can take a picture or video with your phone to confirm you are in the posture you meant to be and rotating on the plane you want to be.

Fine Tuning your Biomechanical Posture, Two-Plane Neutral Arc examples

The examples below allow a variety of fitting solutions for three players, each with a standard 70° lie angle putter. Player A is standing about 6° too tall for his standard putter. Player B has a well aligned posture and lie angle. Player C is bent over about 6° too far for his standard putter.

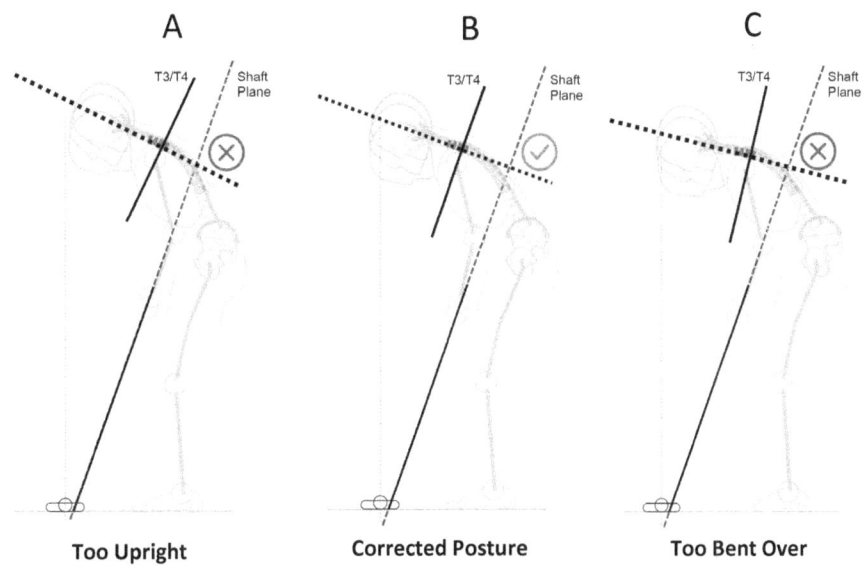

Too Upright   Corrected Posture   Too Bent Over

Players A and C

Each of these players could change posture or lie angle, or change both. Player A could change to a very flat lie angle putter and stay in his original posture, or bend over to match the lie angle of the putter. Player C could change to very upright putter, or stand up taller to match the lie angle of the putter. Our experience is that changing both is a good strategy to try, taking half of the adjustment from the posture and the other half from the lie angle of the putter.

Since it is advantageous to putt in a posture (and with a lie angle) that reduces neutral arc face rotation, bending over a little more, or staying bent over where you started at least, will generally be recommended; as long as you can maintain rotational freedom and your pendulum putting tempo.

Player B

A player already in an aligned two-plane neutral arc posture can experiment with marginally more upright or flatter putter lie angles (and matching posture).

Look at the range of postures and corresponding putter lie angles on the top of page 61.

The standard 70° lie angle putter has evolved to match the neutral arc posture of the average tour player. You don't see many tour players using a 4° upright putter lie angle or a 4° flat putter lie angle.

Putter Length

Lie angle and Putter Length are directly determined by posture. Generally, our experience has been that players have longer putters than they need.
Be careful with the length of your putter. Let your arms hang freely with no tension.

Biomechanical Putting Posture

5.7 The Neutral Putting Arc Face Rotation

The putting stroke is a circle. Its "shadow" (vertical projection of the circle on the ground) is an ellipse. The putter head traces a circle on an inclined plane typically tilted about the 70° lie angle of the standard putter. The putter head will remain square to the circular path, and the arc on the ground, unless rotation is added by the hands/wrists/forearms. The neutral arc face rotation is the putter head arc and rotation <u>square to the arc without any added rotation from the hands/wrists/forearms</u>. The neutral arc face rotation can be calculated by geometry and confirmed by launch monitor metrics.

Neutral Arc Face Rotation
(the calculations are complicated)

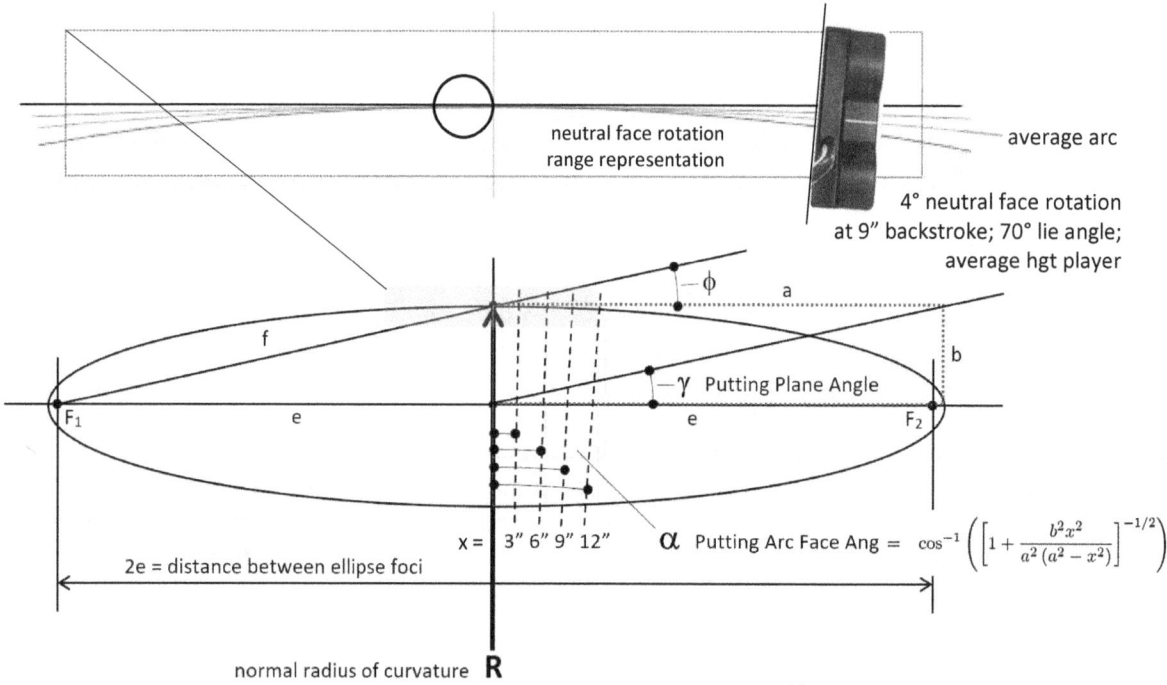

$$\alpha \text{ Putting Arc Face Ang} = \cos^{-1}\left(\left[1+\frac{b^2x^2}{a^2(a^2-x^2)}\right]^{-1/2}\right)$$

The Neutral Putting Arc Geometry

The Putting Arc Face Angle α varies with backstroke length and putting posture. The graph at the end of this section simplifies putting posture inputs which include:

lie angle, height, posture, etc.

The following page is a simplified summary of the Neutral Arc Face Rotation with various alternative putting postures with a 9" backstroke.

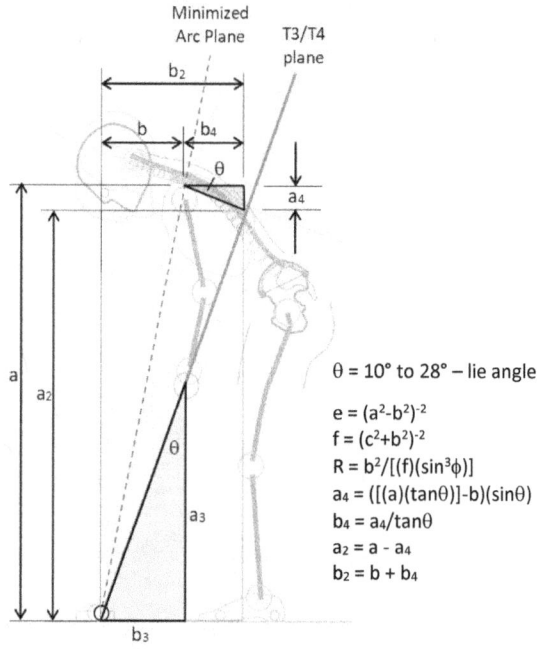

θ = 10° to 28° – lie angle

$e = (a^2-b^2)^{-2}$
$f = (c^2+b^2)^{-2}$
$R = b^2/[(f)(\sin^3\phi)]$
$a_4 = [[(a)(\tan\theta)]-b](\sin\theta)$
$b_4 = a_4/\tan\theta$
$a_2 = a - a_4$
$b_2 = b + b_4$

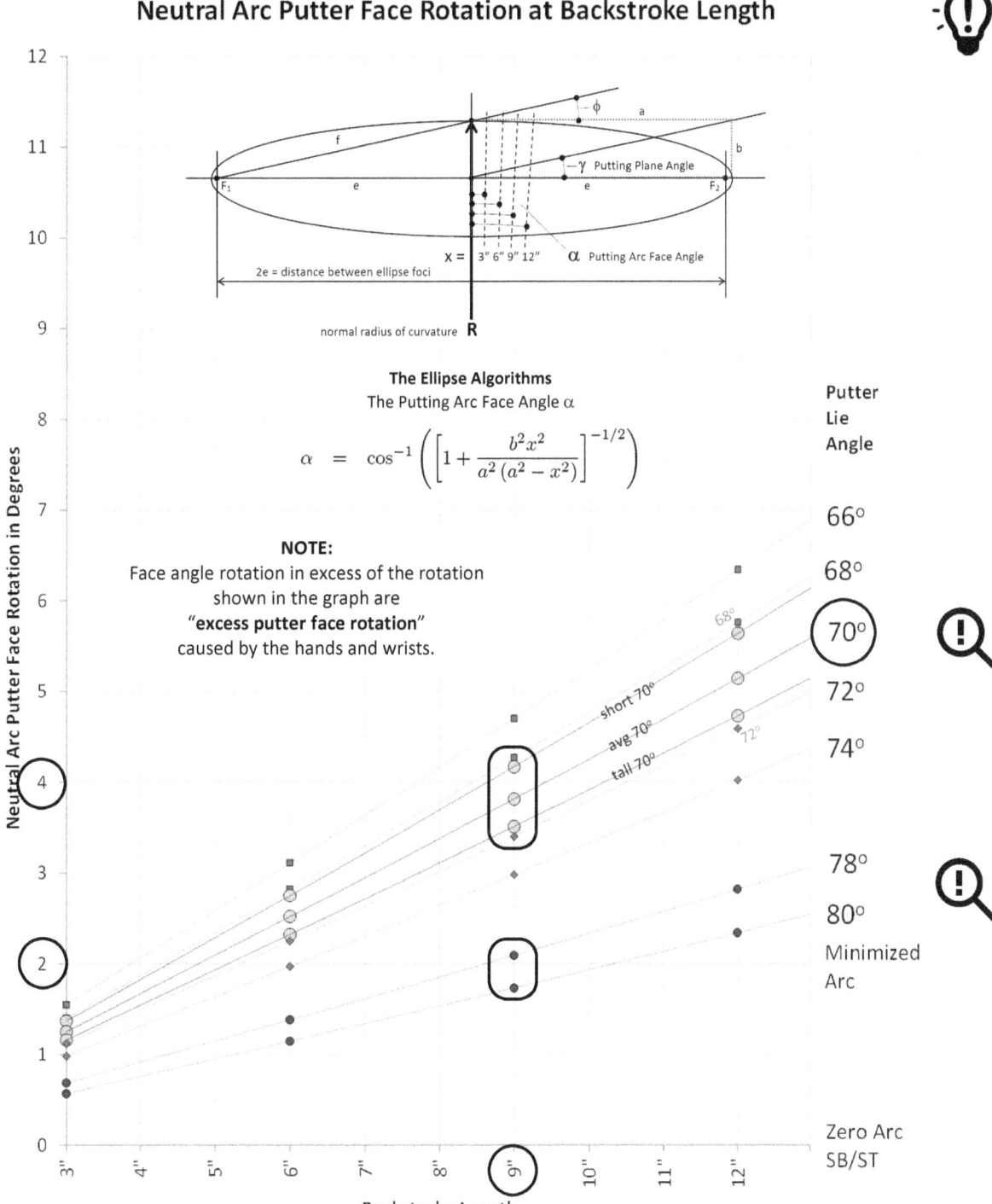

The circles on the chart above indicate the 4° face rotation that is characteristic of a two-plane neutral putting stroke (with no hands/wrists/forearm rotation) resulting from a 70° posture and 70° lie angle putter at 9" backstroke length; and 2° face rotation is typical of minimized arc putting strokes

The data in this graph is calculated from the geometry of the putting arc ellipse. The advent of putting launch monitors allows a player and teachers to have detailed metrics including face rotation during the stroke. This allows a player, with a known posture and lie angle to know the face angle rotation that would result from only the arc and therefore the residual additional arc being introduced by hands/wrists/forearm rotation. This is important information.

Adding Wrist Cock/Release Adds Face Rotation to your Putting Stroke

Adding wrist cock and release to either the two-plane neutral arc stroke or the one-plane minimized arc stroke will dramatically increase face rotation.

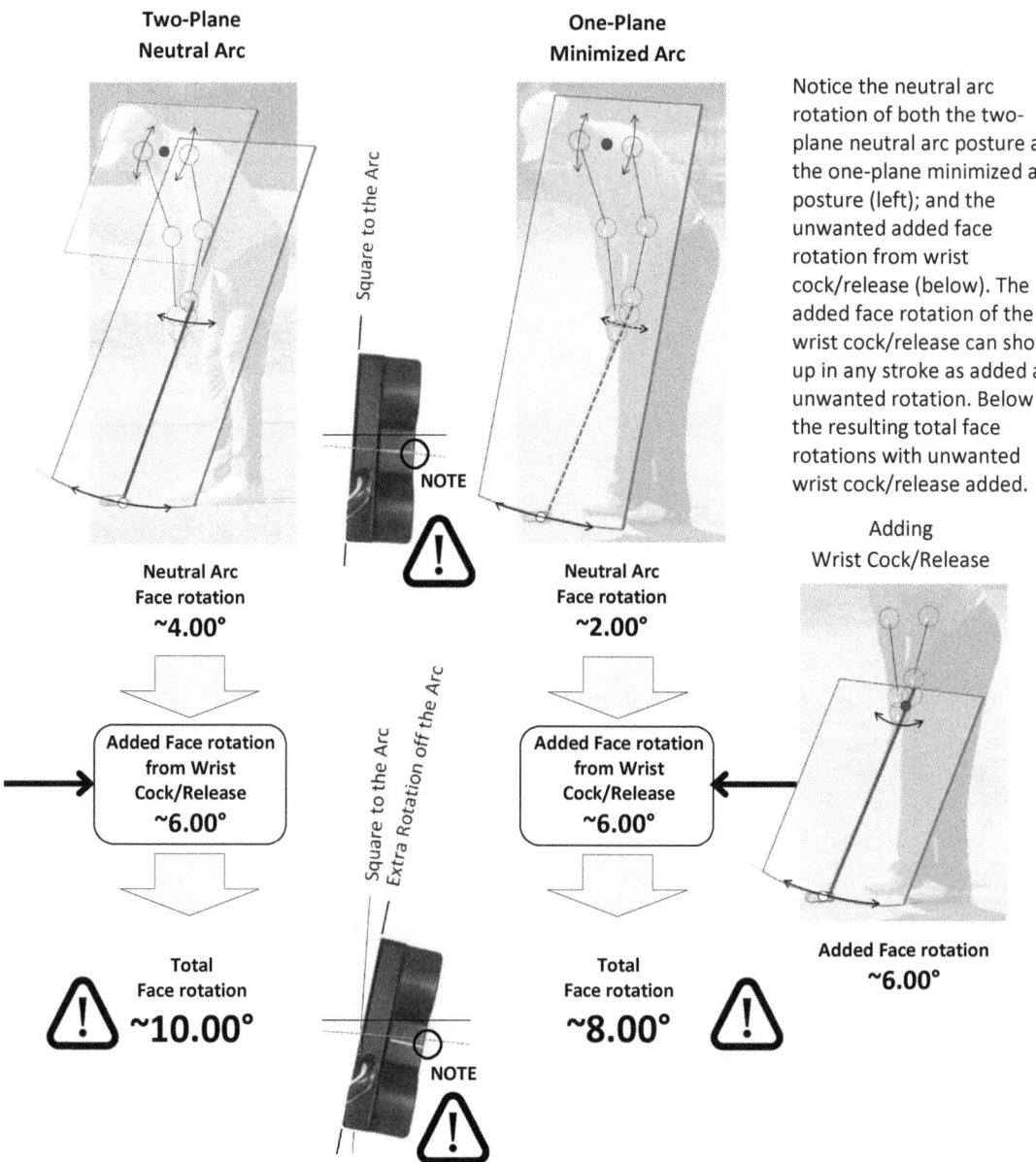

The extreme rotations (caused by added wrist cock/release) are very common among amateurs.

Unwanted wrist cock/release can play havoc with face rotation. Amateurs routinely add as much as 6° of face rotation by adding unwanted hand and wrist action into their stroke; creating total rotations of 8° to 10° or even more, sometimes much more. The growing use of putting launch monitors will make this clear. The graph on the previous page shows the neutral rotation for a full range of postures.

The neutral putting arc for your personal putting posture and biomechanical arc can be used with putting launch monitors to develop a more efficient putting stroke with the correct arc and face rotation.

Biomechanical Putting Posture
5.8 Measuring Spine Angle Accurately is Important

This part of the discussion could easily have been in the "Myths and Misunderstandings" section, but we decided to have this discussion now.

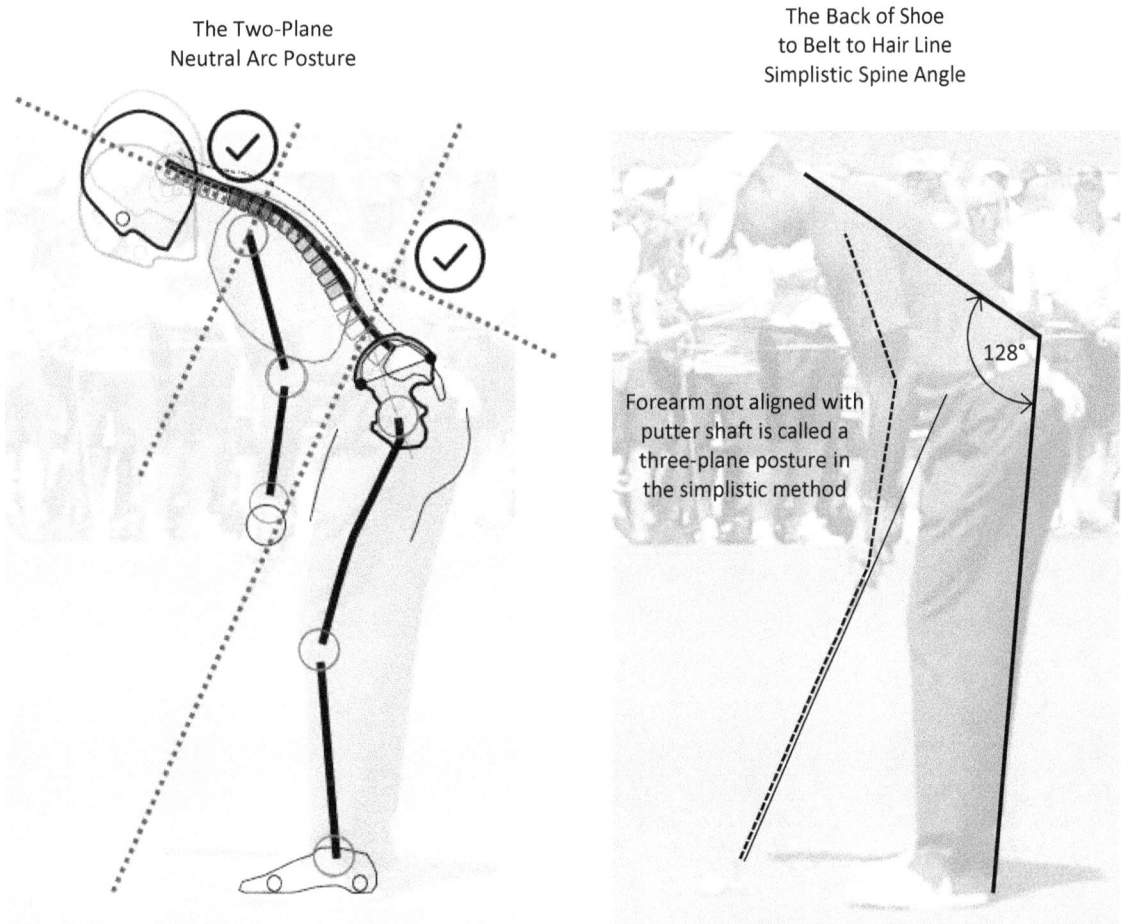

The "simplistic" spine angle method users often argue that this "spine angle" should be between 111° and 118° which would ask Tiger Woods (and most PGA tour players) to bend over. A lot more.
All of the tour players in the Two-Plane Neutral Arc Posture (about 95% of the tour) have simplified "spine angles" between 120° and 130°.

The simplistic method is quick and easy; but simply too crude to be useful for detailed analysis of putting posture. The simplistic method ignores pelvic tilt, lumbar lordosis and thoracic kyphosis.

The overlay method we have used is time consuming, even somewhat tedious, but it yields very useful information about the actual position of the upper thoracic spine and the center of putting stroke rotation at the T3/T4 joint. The development of new software to allow quick, near real time, overlay of detailed biomechanical information on video images is needed.

One last point. The idea that the upper arm, forearm and putter shaft alignment creates a three-plane posture with more moving parts is absurd. There are no more moving parts whether the forearm is aligned with the putting shaft or not.

Section 6
Moment of Inertia (MOI) and STABILITY

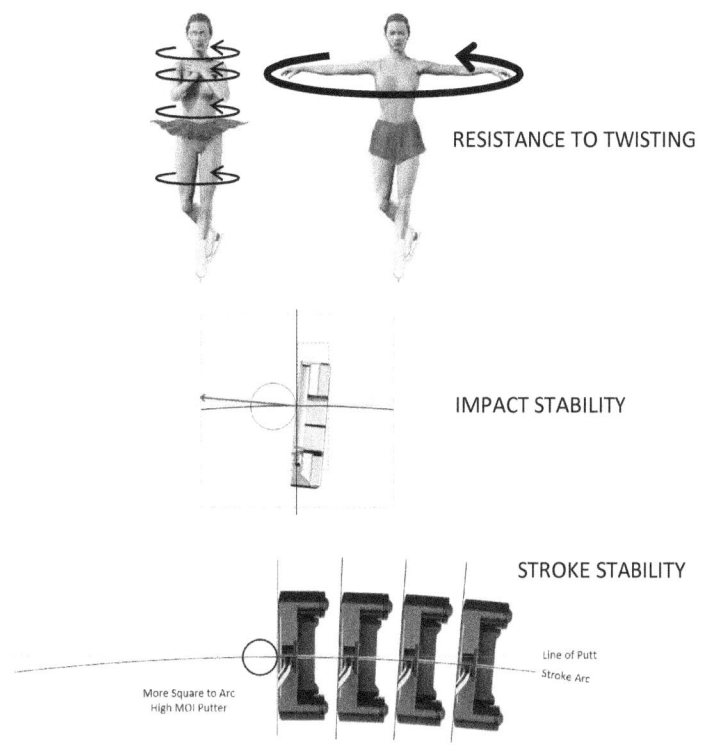

Moment of Inertia (MOI) and STABILITY
6.1 *Putter Stability vs Putting Stroke Stability*

We are going to look at the two most important kinds of stability in putting. The first, impact stability, everyone knows a little about. The second, stroke stability, is understood only generally or not at all. The magnitude of improvement with MOI is stunning. The idea of weight optimization is entirely new.

Impact Stability (Forgiveness)
(twist at impact from an off-center hit from lack of MOI)

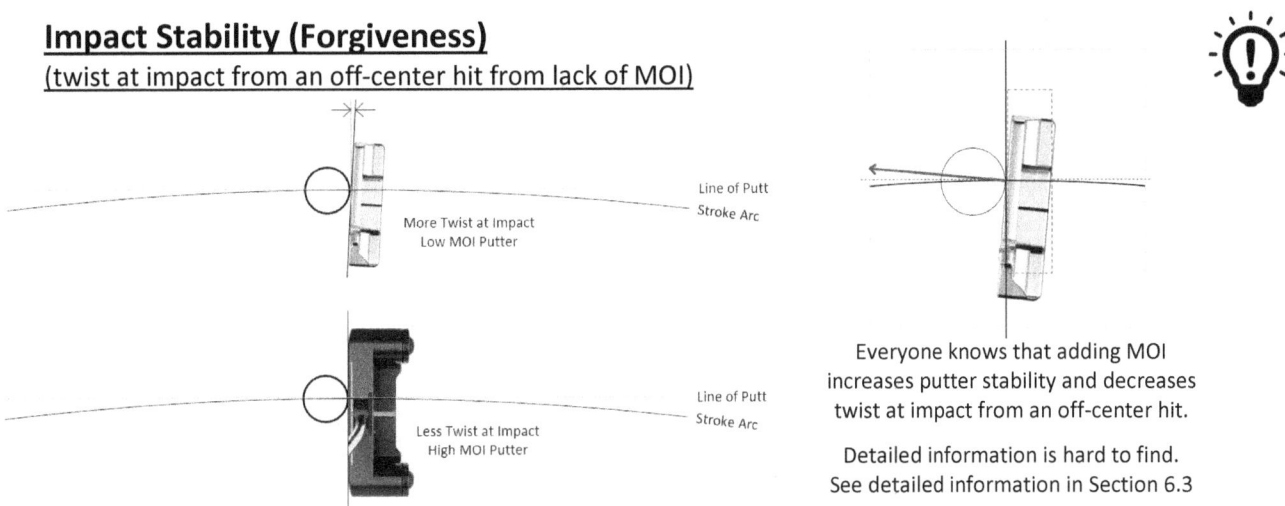

Everyone knows that adding MOI increases putter stability and decreases twist at impact from an off-center hit.

Detailed information is hard to find. See detailed information in Section 6.3

Adding MOI reduces twist at impact and the distance lost to an off-center hit, dramatically.

Stroke Stability
(MOI resists twisting in the stroke itself)

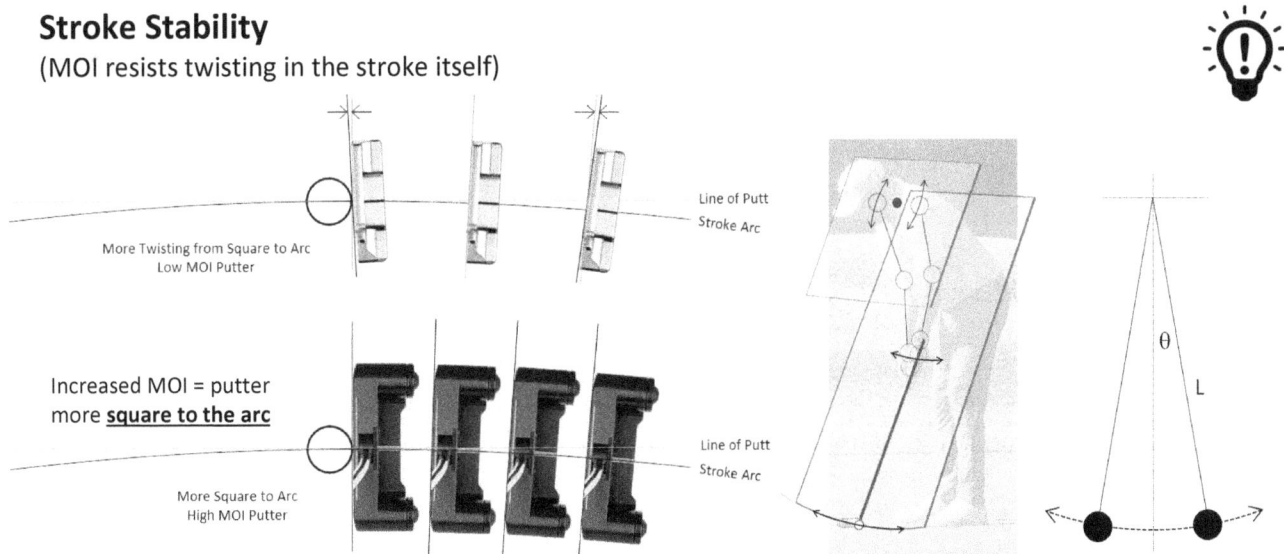

Face angle can be measured at any point in the stroke by a putting launch monitor; but the most important from a putting results perspective is at impact. Tests have demonstrated that face angle variation at impact can be reduced dramatically by adding MOI. These tests have demonstrated that adding weight/MOI reduces face angle variation only up to the individual's optimum weight. This is a new discovery that is explained by the physics of the personal effective pendulum.
This is a HUGE discovery of game changing importance.

A quick review of the concepts of moment of inertia (MOI) will help lay the foundation.

Moment of Inertia (MOI) and STABILITY

6.2 Moment of Inertia is **STABILITY**, Measured

Manufacturers have continued to treat MOI as a proprietary detail; often refusing to disclose the actual MOI of their putters. They still claim to have increased stability. This is nonsense. Marketing gibberish.

MOI and Stability are scientifically identical

From a purely scientific perspective, Moment of Inertia (MOI) and Stability are not just related; they are identical, synonymous. MOI and Stability are <u>exactly</u> the same. Test results confirm that twist at impact from an off-center hit is directly related to MOI. Higher Stability = Higher MOI. Period.

MOI (Moment of Inertia) is resistance to twisting.

$$I = md^2$$

The formula is $I = md^2$

Moment of Inertia (I) = mass (m) times distance (d) squared; distance (d) is the distance from the COG (center of gravity).

A Tale of Two Putters
both with 350 g head weight.

A 350 gram putter with a lot of its mass at its COG will have a low MOI.
A 350 gram putter with a lot of mass far away from its COG will have a higher MOI.
The <u>**ONLY**</u> way to increase the MOI of a 350 gram putter is to move its mass further from the COG.

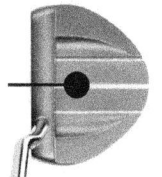 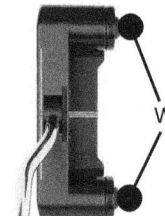

Weight near the center of gravity <u>decreases</u> MOI

Weight away from the center of gravity <u>increases</u> MOI

Traditional Mid-Mallet Weight Distribution 350 g	High MOI Mallet Weight Distribution 350 g
Less STABLE **4,164** gcm² MOI	**6,900** gcm² MOI More STABLE

The Radius of Gyration ($\sqrt{I/m}$), the square root of MOI divided by mass,
is the effective distance of the total putter mass from its COG.

The Radius of Gyration of the high MOI mallet (4.44") is only a little larger than the V-Line (3.44") but this creates a lot more resistance to twisting, just like a figure skater; who spins faster with a smaller radius of gyration, and slower with a larger radius of gyration.

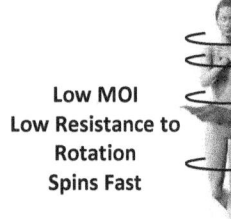 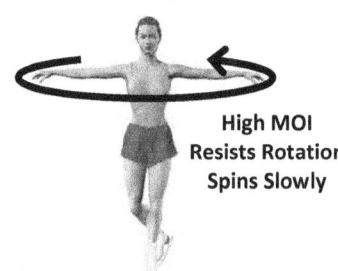

Low MOI
Low Resistance to Rotation
Spins Fast

High MOI
Resists Rotation
Spins Slowly

You should demand to know the MOI of your putter because MOI is the scientific measure of its stability. You want a more STABLE putter. MOI is STABILITY.

Moment of Inertia (MOI) and STABILITY

6.3 Impact Stability (Forgiveness)

Marketing departments of major putter companies continue to describe stability as if it is independent of MOI. It is not. Any increase in stability will be reflected in increased MOI (as a scientific necessity). Rigorous testing with very accurate putting launch monitors demonstrates that increased MOI stabilizes the putter in an off-center hit.

The direct measure of a putter's actual scientific stability is moment of inertia (MOI).

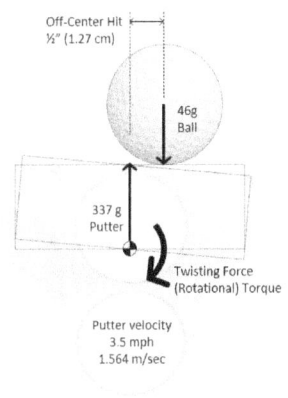

MOI is Resistance to Twisting

The **Torque** (35,900 gcm) applied to the putter by an off-center hit is resisted by its **MOI** (3553 gcm^2). The resulting twist at impact was about 0.70° in the testing for the original 2 Ball putter shown here.

MOI is Stability

Claims of increased putter head stability with no increase in MOI is marketing gibberish.

Torque (35,900 gcm)

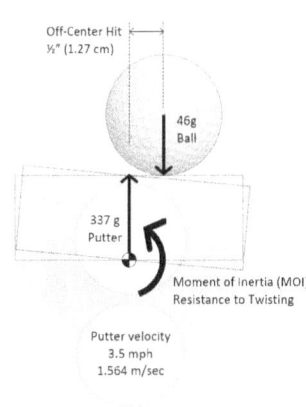

MOI (3553 gcm^2)

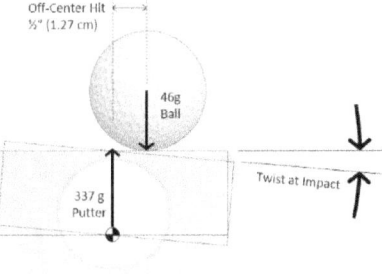

Twist at Impact creates two undesirable effects:

1) direction error (directly from the twist);
2) distance error (because energy is consumed by the twisting).

The ball starts off line and doesn't roll as far.

The putter twists around its center of gravity first, especially during the impact period. A very short time; about 0.00043 seconds. A little less than one half of one thousandth of a second. The putter continues to twist after the ball leaves the face. The total twist is more than three times the twist while the ball is on the face.

<u>The twist at impact with a robot is LESS than the twist in the hands of a human</u>. The robot has a grip of steel. The human hand is softer. The human hand may be very much softer if the player's grip is softer.

Detailed Twist at Impact from an Off-Center Hit Test Data
Test data confirms that stability and MOI are directly related

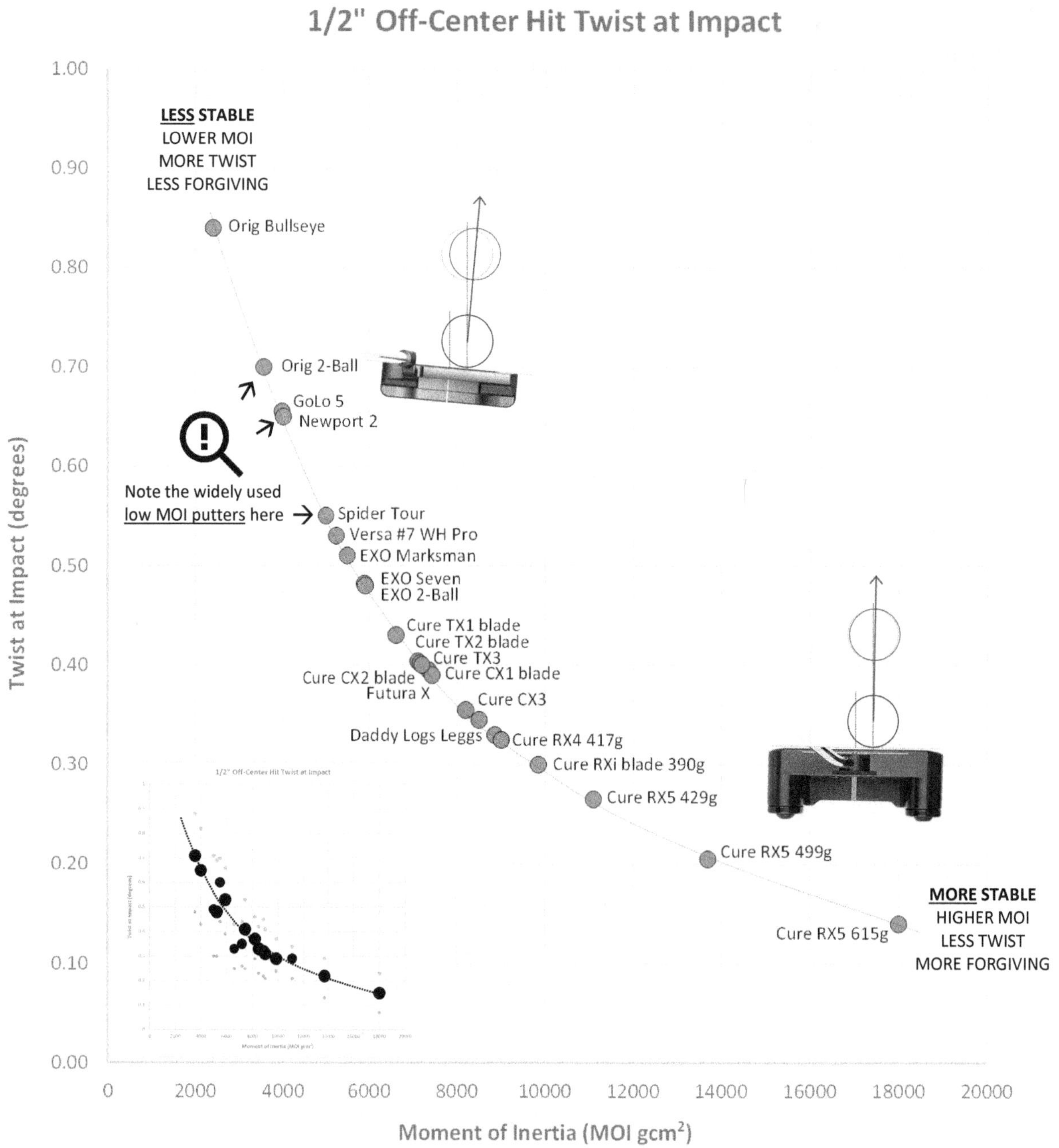

Presented in the graph above are the results of robot testing with various well-known putters on a Quintic Ball Roll Systems putting launch monitor with an Iron Archie Putting Robot. A minimum of 20 putts was performed with every putter. The inset (bottom left of the graph above) shows all of the actual test results with the visually fit curve overlaid as a dotted line. The larger graph above shows only the final curve with each of the putters labeled.

The test results were clear. Every increase in MOI increases stability no matter where you are on the curve. Many manufacturers claim increased stability without disclosing the actual increase in MOI. Any claim of increased stability that is not reflected in MOI is marketing gibberish. They claim increased stability but refuse to release their MOI data. Why? The most likely answer is the data does not support their claim of increased stability. The advent of putting launch monitors will bring this dissembling to an end, eventually.

A Tale of Two Putters' Distance Control

The energy consumed in twisting is reflected in a decrease in impact ratio. Both the Newport 2 and the RX5 had an impact ratio of 1.70 on a center hit. The Newport 2 impact ratio fell from 1.70 to 1.53 on a ½" off-center hit, a 10% reduction in ball speed (0.17/1.70 = 10.0%). The RX5 fell to from 1.70 to 1.69 on a ½" off-center hit, a 0.50% reduction in ball speed (0.01/1.70 = 0.50%).

A 30' putt with the Newport 2 will be off line and come up over **3 FEET** short.
A 30' putt with an RX5 will be on a better line and be shortened by only **2 INCHES**

The combination of launch direction change and distance erosion results in very poor performance for low MOI putters in comparison to the higher MOI putters. Every putter has a discreet and knowable MOI which places it somewhere on the line of the graph above. The higher MOI putters to the right perform better. They are more stable. This is physics.

A Dramatized Illustration of Stabilized Distance and Direction
The Holy Grail of Putting

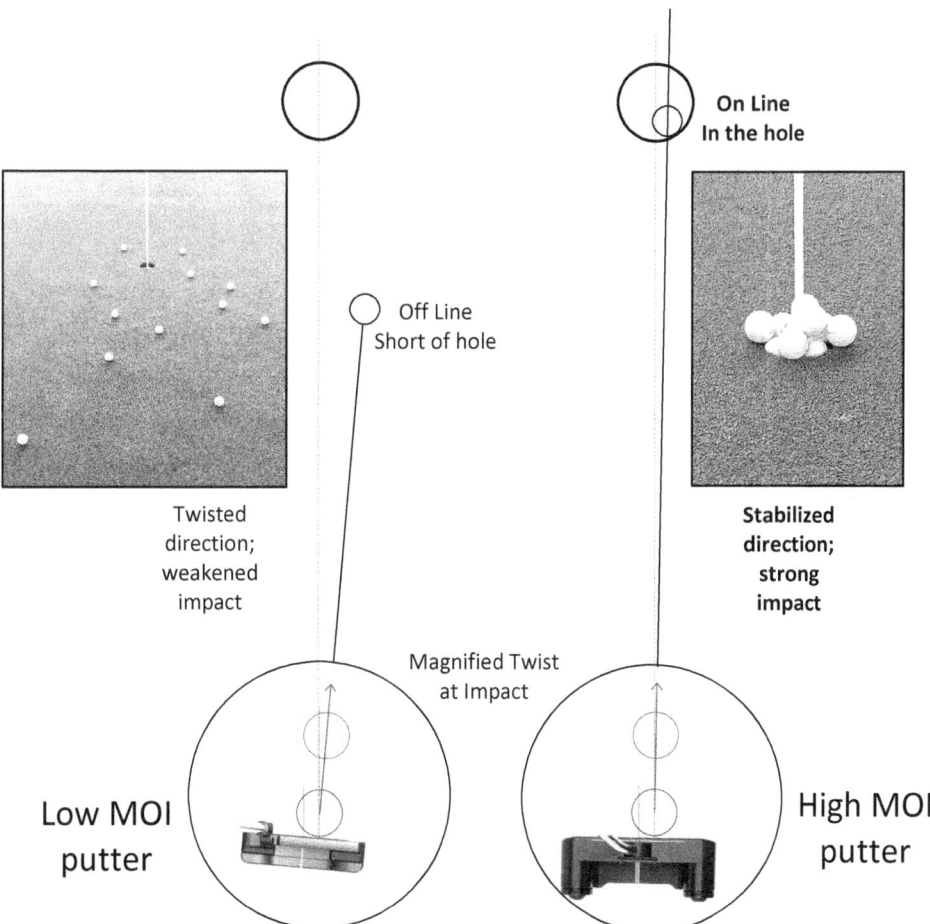

The use of a high MOI putter will create a much tighter putt dispersion. Really.

Moment of Inertia (MOI) and STABILITY
6.4 Weight Optimization Fitting and Pendulum Physics

The appearance of a personal optimum weight in the human testing for face angle variation at impact was a BIG surprise. We would not have been surprised if the improvement had continued to the right in both of the graphs below; like it did with additional MOI in stability at Impact experiments (top graph).

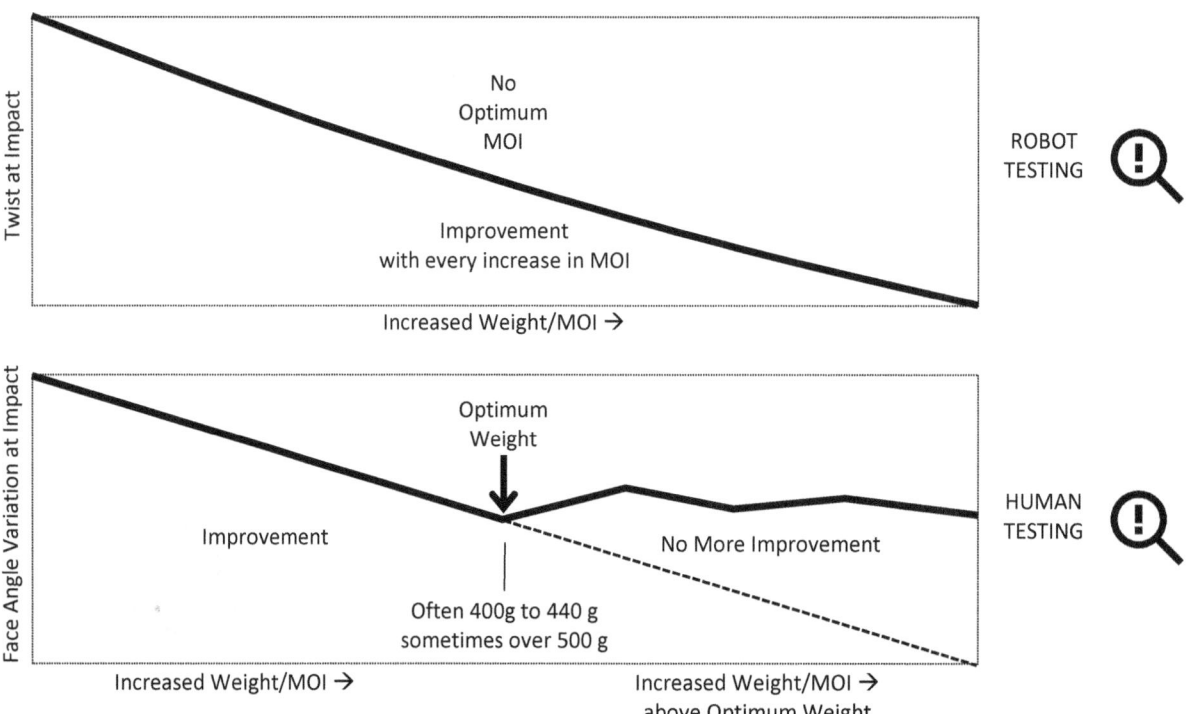

The scientific explanation of Weight Optimization Fitting almost certainly involves pendulum physics. The existence of an optimum weight for every individual may be explainable as **fine tuning your personal second harmonic**.

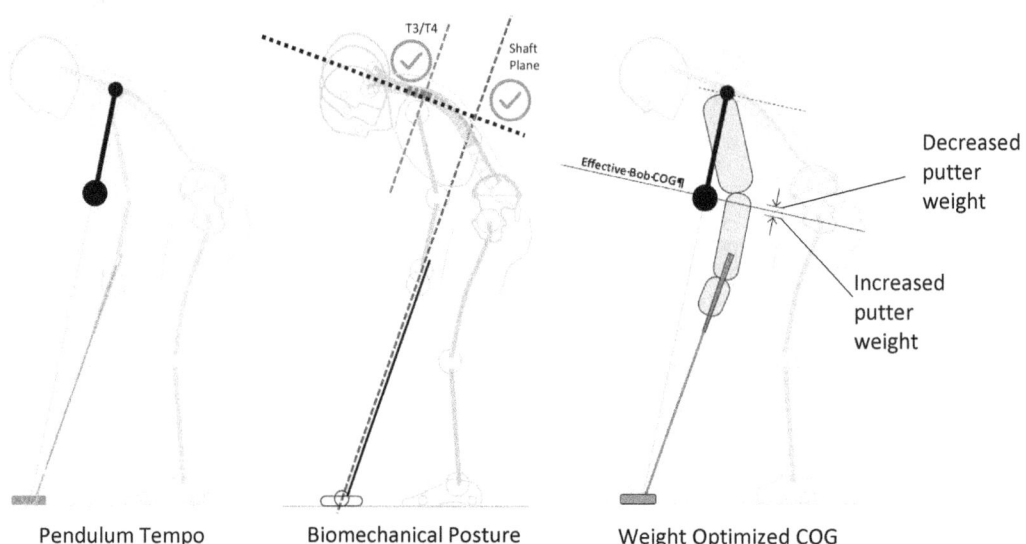

The second graph above previews the weight optimization test results in the next section. The figure to right above explains the context of weight optimization in pendulum physics.

Moment of Inertia (MOI) and STABILITY

6.5 Stroke Stabilization
Face Angle Variation Range Reduction

Stroke Stability
(MOI resists twisting in the stroke itself)

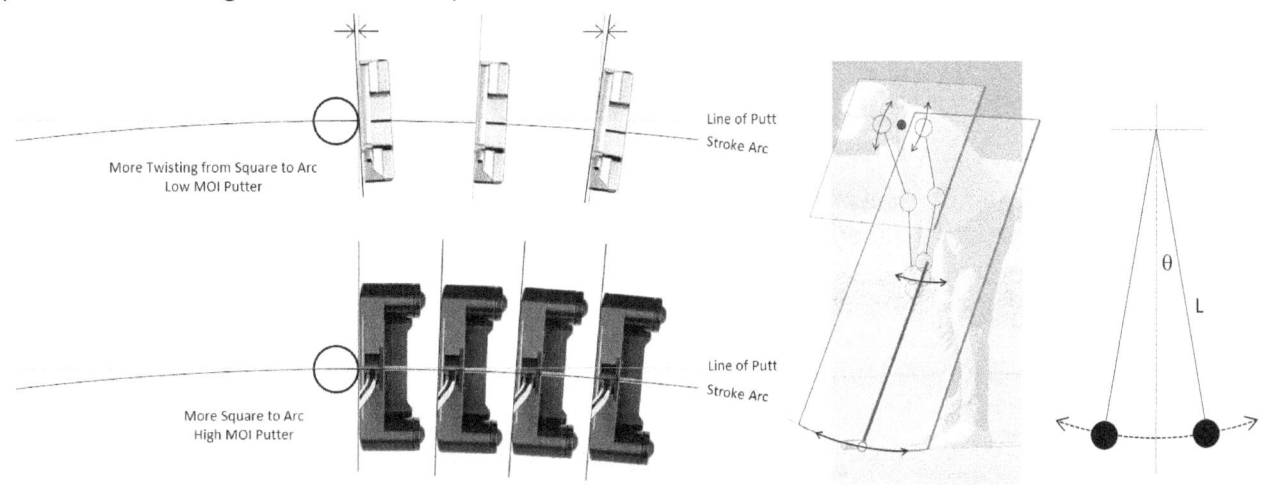

Tests have demonstrated that adding weight/MOI reduces face angle variation <u>only</u> up to the individual's optimum weight; adding weight beyond this point does not result in improvement.

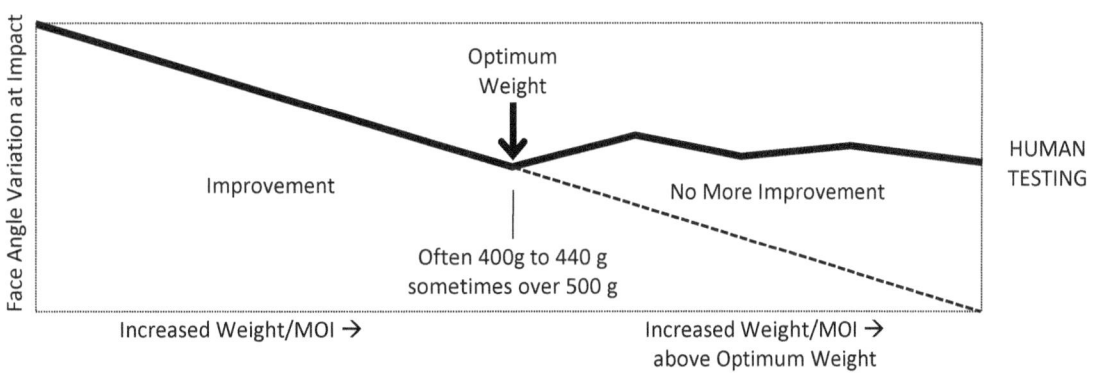

The existence of the optimum weight would never have been discovered using the limited weight range of most putter manufacturers. All tests with limited weight range putters would have been within the "improvement" segment (left above). Our initial testing used a Cure Rx Series putter (right below) with a far higher than standard weight range. This allowed us to make the surprising discovery of Optimum Weight, generally a lot higher than standard putters.

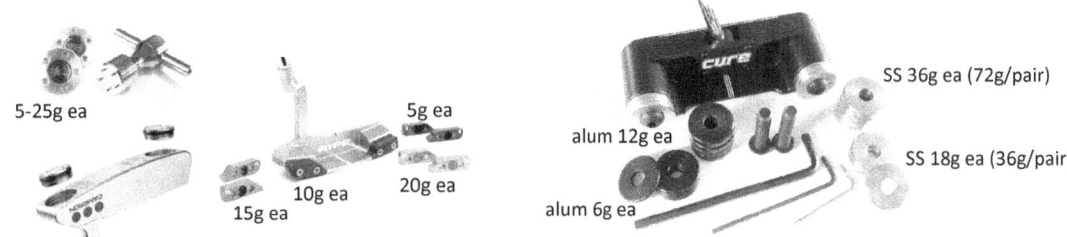

Limited Weight Range
30g-40g range (335g to 375+/-g)

Extreme Weight Range
200+g range (350g to 550+g)

Weight Optimization was based on <u>Extreme Weight Range</u> Testing

One-Putt Enterprises (the Quintic Ball Roll Lab in the USA) Massachusetts tested the extreme weight range of the Cure RX Series putter. In this testing we discovered the surprising benefit of weight optimization fitting. Higher than standard weights are required.

The testing proved that face angle variation range at impact improved dramatically as weight was added (increasing MOI). The improvement eventually ended or deteriorated at a particular weight, different for every player. The optimum weight for players has ranged from 375 g to 535 g. <u>Both the extreme weight required and concept of an individual optimum weight were a surprise.</u>

The Individual's Optimum Weight is clear in the test results:

The alpha test was with a 16 handicap golfer; an average putter for his handicap with a 2.9° face angle variation range in his baseline putter test. 9 Cure RX Series putter configurations were tested. Face angle variation range decreased (improved) in the first 5 tests; and then failed to improve more in the last 4 tests. The putters in tests 1-5 "felt" good to the player; the putters in tests 6-9 "felt" too heavy to the player. An OPTIMUM WEIGHT was discovered at 409 grams for this player.

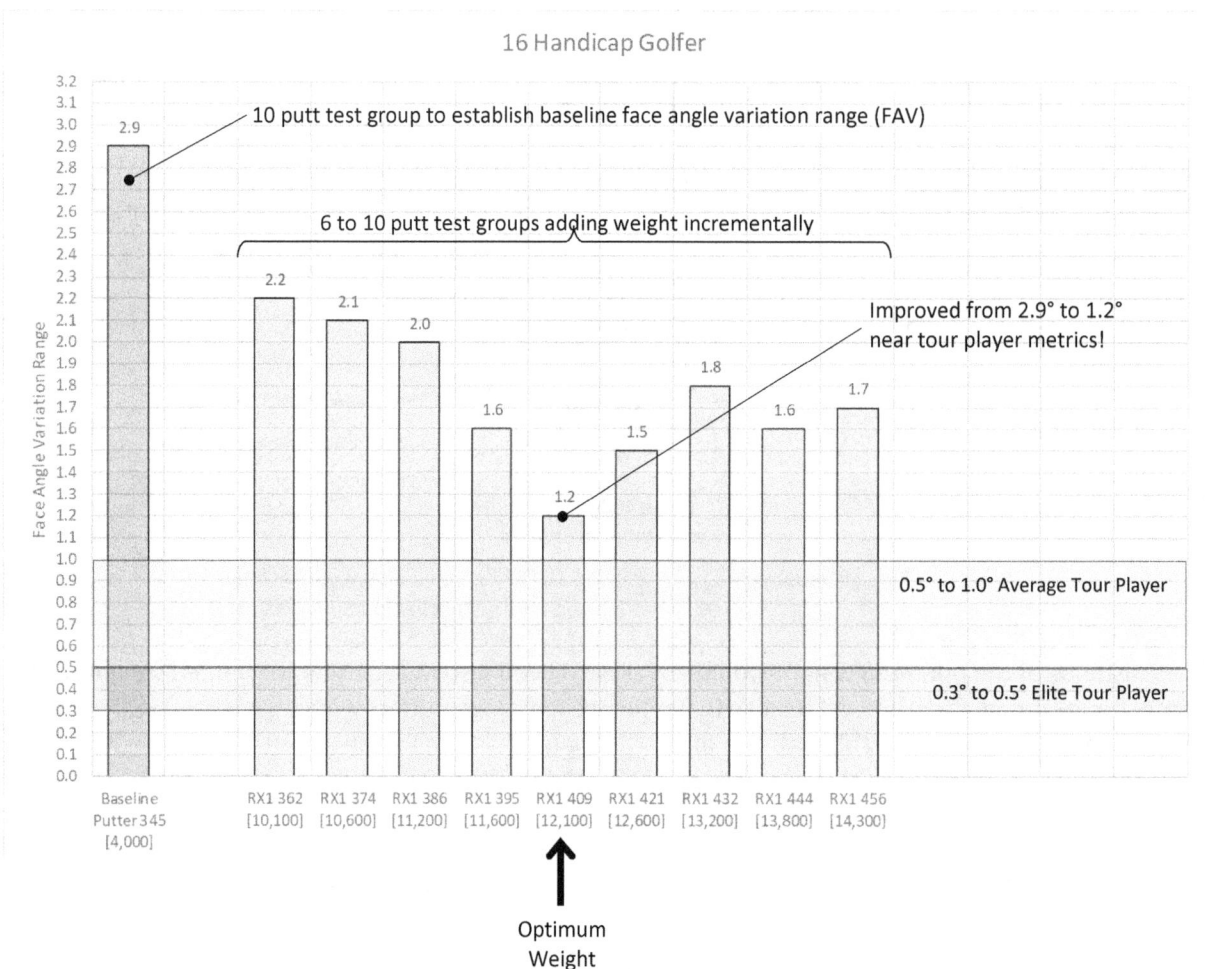

The alpha test discovered an optimum weight outside the industry expectation (as indicated by the average weight of available putters) of between 345g and 355g. Improvement in a number of putter launch metrics was noted, including importantly: putter speed variation range and launch angle range.

Dramatic FAV Improvement up to Optimum Weight in the test results:

The beta test was with a 6 handicap golfer; with a 2.1° face angle variation range in his baseline putter test. 9 Cure RX Series putter configurations were tested. Face angle variation range decreased (improved) in the first 4 tests; and then failed to improve more in the last 5 tests. The putters in tests 1-4 "felt" good to the player; the putters in tests 5-9 "felt" too heavy to the player. An OPTIMUM WEIGHT was discovered at 395 grams for this player.

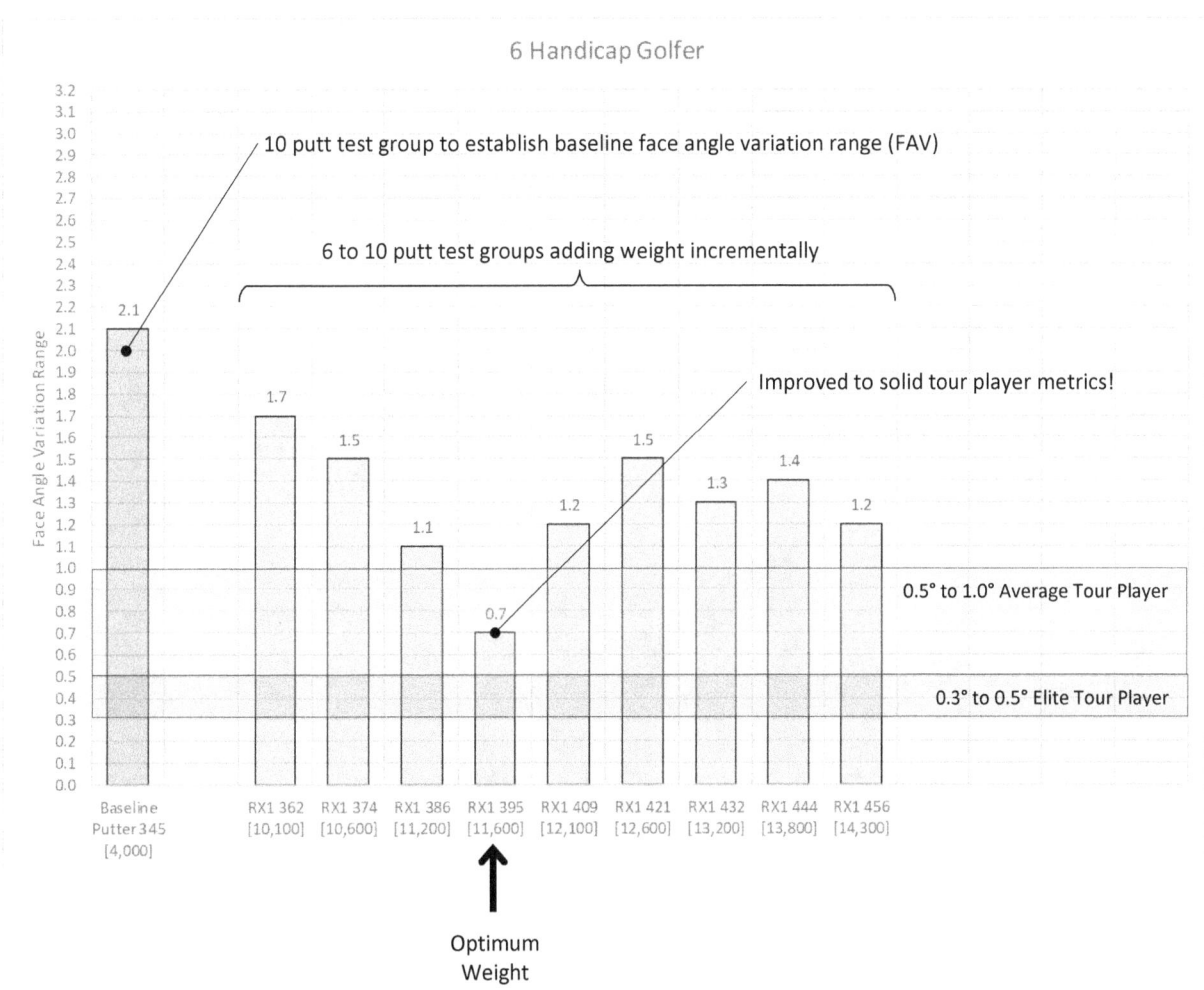

The optimum weight for face angle variation range and putter speed variation range coincided at the same weights; with the same pattern: improvement up to the optimum weight and then no further improvement.

Subsequent tests have been done with players of all skill levels including PGA Tour players, PGA Tour Champions players, LPGA Tour players, web.com and mini Tour players, International Tour players and amateurs from +4 handicap to -32 handicap. The results persistently confirm an optimum weight for players of all skill levels a putter weight heavier than generally available in the market.

This is scientific proof that each player has an optimum weight, individually determinable.

The Face Angle Variation (FAV) results are GAME CHANGING!

A weight optimized, MOI maximized putter is proven to improve face angle variation dramatically. The test results summarized below are not unusual. Improvement on the order of or excess of 50% reduction in face angle variation is normal. This is the magnitude of improvement we have seen for players of all levels from weekend warrior to tour players.

The Chart below shows actual test data from players ranging from 16hcp to Tour players. Not only did every player experience dramatic improvement, most players reached Tour level consistency simply by using a Cure at their ideal weight.

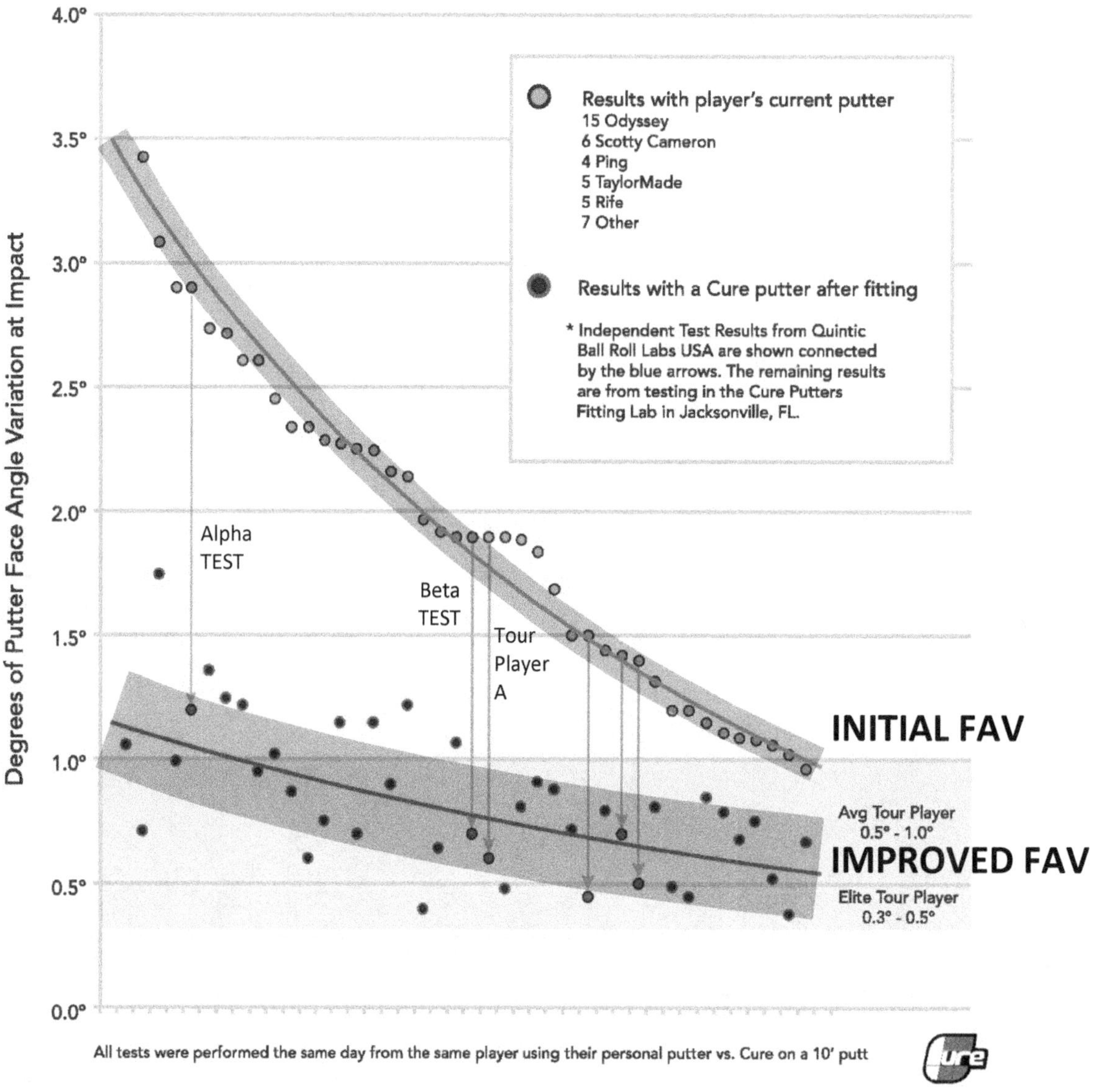

○ Results with player's current putter
- 15 Odyssey
- 6 Scotty Cameron
- 4 Ping
- 5 TaylorMade
- 5 Rife
- 7 Other

● Results with a Cure putter after fitting

* Independent Test Results from Quintic Ball Roll Labs USA are shown connected by the blue arrows. The remaining results are from testing in the Cure Putters Fitting Lab in Jacksonville, FL.

INITIAL FAV

Avg Tour Player 0.5° - 1.0°

IMPROVED FAV

Elite Tour Player 0.3° - 0.5°

All tests were performed the same day from the same player using their personal putter vs. Cure on a 10' putt

Note that the ALL players improved their FAV performance, regardless of skill level. Most players improved into the shaded area at the bottom of the graph above representing the face angle variation of Tour Players. The average tour player has a face angle variation range of from 0.5° to 1.0°. The elite tour player has a face angle variation range of 0.3° to 0.5°.

MOST of the players above improved from average to poor into or close to tour caliber. The meaning of this is clear. Average players can improve their putting to tour caliber without a putting lesson. Simply put a weight optimized, MOI maximized putter in their hands and their putting launch monitor metrics improve, a lot! Let that sink in. If you want to be a better putter, weight optimize your putter.

Note "Tour Player A" in the graph above. This was Rod Spittle, PGA Tour Champions, tested in December 2014 at the PGA Tour Academy Facility at TPC Sawgrass in Ponte Vedra FL. The test was conducted by Jim MacKay (One-Putt Enterprises, the Quintic Ball Roll Lab in the USA) with Steve Davis (author). Rod was using a 350 g putter with less than 4000 gcm^2 MOI.

Weight optimization testing increased Rod's putter weight to 535 g. His face angle variation decreased from a very poor 1.85° to a very good 0.65°. In the next three years his tour stats improved dramatically. He reduced his putts per green in regulation from 1.830 putts/GIR to 1.753 putts/GIR. Hitting 75.13% of the greens in regulation (Rod's average) this represents an improvement of over one stroke per round. Rod's scoring average has improved from 71.74 to 69.76 strokes. These improvements with a weight optimized high MOI putter resulted in a substantial improvement in his average finish position.

Note in the results also that the average improvement in face angle variation range is about 50%. This is a HUGE, game changing, shout from the rooftops, improvement.

With a **50% improvement** in face angle variation range the **hole feels twice as big**!

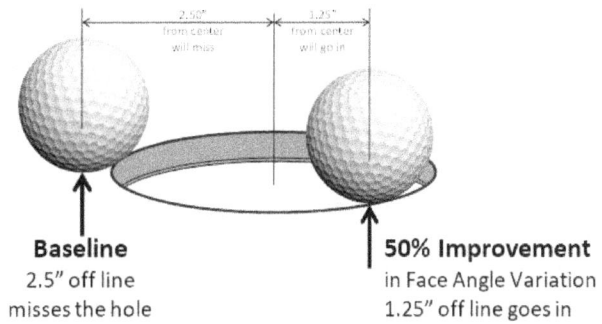

Baseline
2.5" off line
misses the hole

50% Improvement
in Face Angle Variation
1.25" off line goes in

Moment of Inertia (MOI) and STABILITY
6.6 MOI Summary of Findings
Two Kinds of Stability

Impact Stability/Forgiveness

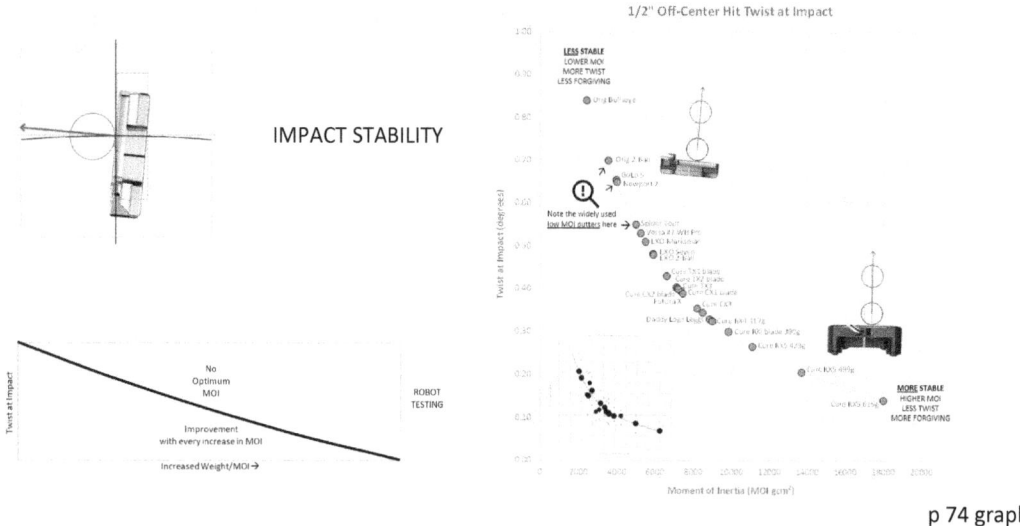

p 74 graph

Every increase in MOI improves stability and forgiveness.

Stroke Stability (Face Angle Variation Reduction)

p 80 graph

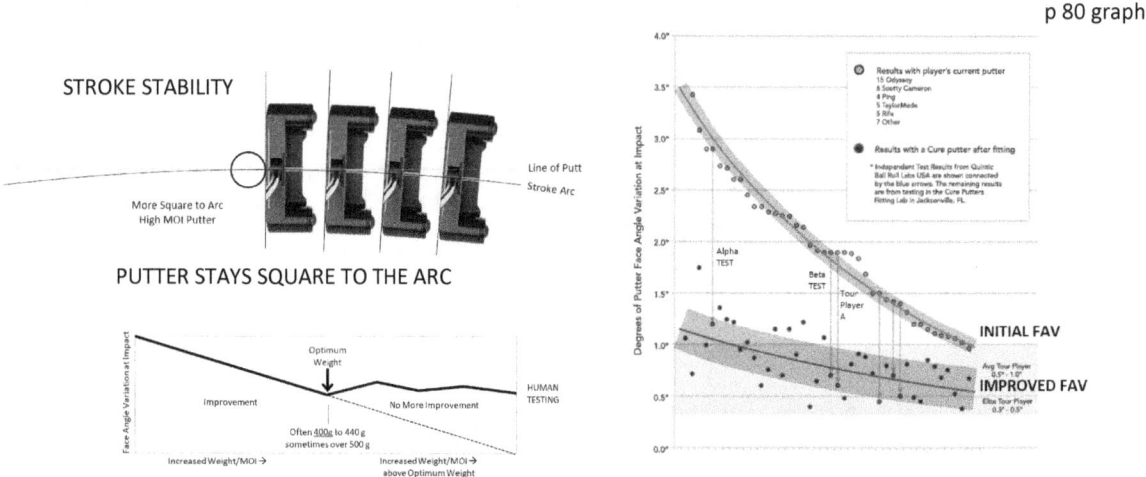

Improvement with added MOI only up to individual player's optimum weight. Dramatic improvement. 50% (or more) reduction in face angle variation at impact.
MIND BENDING IMPROVEMENT.   GAME CHANGING IMPROVEMENT.

Go back and look at the detailed data in the graphs on pages 84 and 90 above.

NOTE:

These two types of stability will drive the design of putters in the future. Real data will make it impossible for putter manufacturers to continue claiming improved stability without real and material increases in MOI. Weight optimization fitting will force manufacturers to provide much heavier putters, for some players, and much wider weight range (375 g to well over 500 g) for proper fitting.

Moment of Inertia (MOI) and STABILITY
6.7 Weight Optimization Protocol

Weight Optimization Protocol Outline

Determine the Characteristics of the Player's Current Putter:
Document the make, model and characteristics of the player's current putter:
1. Make and Model
2. Type (blade, small mallet, large mallet, high MOI mallet)
3. Weight, estimated MOI, Loft and Lie

Baseline Test:
1. Test the player with their current putter with a group of from 6 to 10 putts.
2. Record the results on the Weight Optimization Results form.

Weight Optimization Testing:
Test at a wide range of weight increments until optimum weight is indicated:
1. Add weight incrementally (increasing the MOI with each weight increment)
2. Continue with 2 or 3 tests above indicated weight optimum (to confirm)
3. Record the results for each weight on the Weight Optimization Results form

Face angle variation range will improve with each added weight increment until the player's optimum weight is reached. After the optimum weight, incremental weight increases no longer improve face angle variation range. Testing should focus primarily on face angle variation (**directional control**) and secondarily on ball speed variation (**distance control**); then thirdly on launch, spin and other launch monitor metrics. Note the skew of miss maximum left and right.

In the schematic example below, the player's optimum weight proved to be about 434 grams. Testing in our lab with many players of all skill levels has demonstrated that the optimum weight for most players is in the range of 400 to 440 grams, with many players reaching their optimum at over 500 grams. Blast Motion lab testing indicates that the general optimum weight may be at around 500 grams.

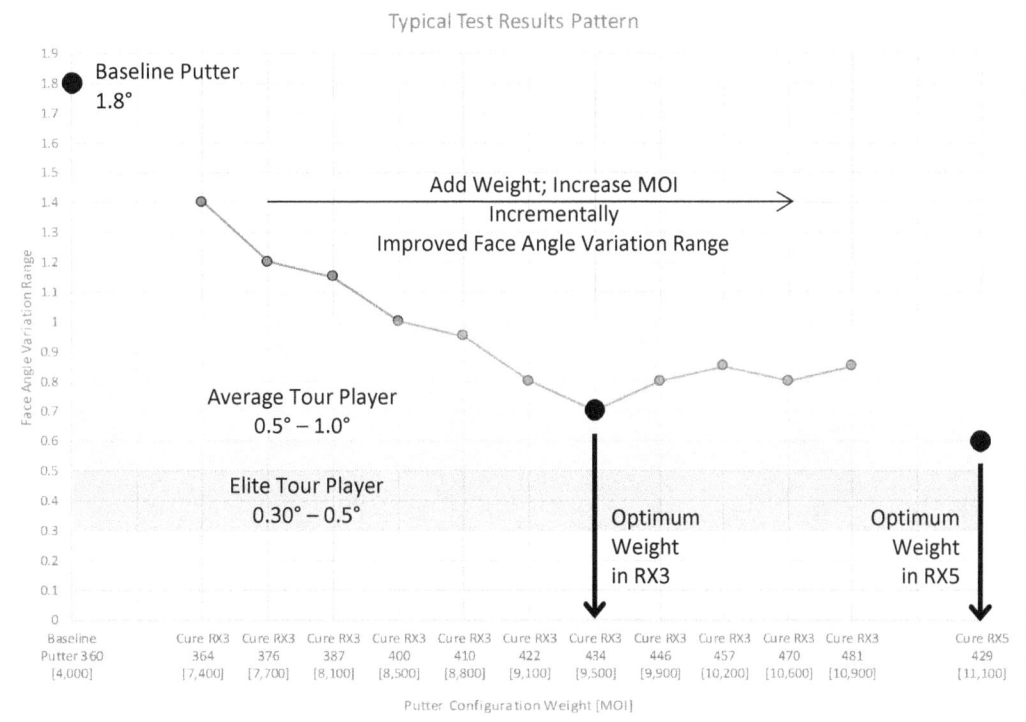

Dynamic Fitting

Weight optimization fitting is a dynamic fitting protocol. The process of putter **weight optimization** is based on the discovery in wide weight range testing that each individual has an **optimum putter head weight**. This weight optimization process finds the putter head weight at which face angle variation is reduced to the smallest range possible while maintaining a player's tempo and natural putting stroke.

The weight optimization process works by starting at a putter head weight similar to the player's existing putter and then adding weight to the head (increasing the MOI of the putter head) in small increments until a weight is reached that is clearly too heavy for the player.

> NOTE: This can be done without any putting performance data (on the green outdoors, or indoors on an artificial surface, even carpet). Experience has determined that finding the first weight that the player reports is simply too heavy is very close to the player's optimal weight when launch monitor performance data is used to measure face angle variation.

Most traditional putters are not weight adjustable; those that are adjustable have a very limited range of adjustability of only a few grams (maybe as much as 50 grams). Cure RX Series putters have a weight adjustment range of well over 200 grams (in about 12 gram increments).

The extreme weight adjustment range of the Cure RX Series putter design that allowed the discovery of Weight Optimization Fitting and are still the best putters to use in this process because of their extreme weight range.

Weight Optimization Fitting Record

Player Name: _____

Handicap _____ Date: _____

Putter	Weight (grams)	MOI (gcm^2)	Face Angle variation range	Max Left	Max Right	Ball Speed variation range	Twist at Impact variation range	Launch Angle variation range	Zero Skid (in) variation range	Zero Skid (in) variation range	Spin Notes

Baseline

model											

Length _____ Lie _____ Loft _____ Other notes: _____

Weight Optimization Tests

model											
model											
model											
model											
model											
model											
model											

Section 7
Green Speeds, the Modern Putting Stroke and the Modern Putter

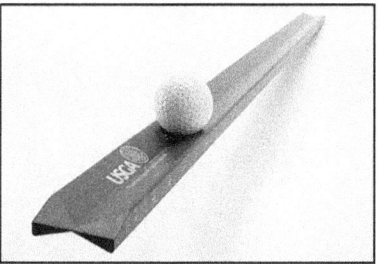

Green Speeds, the Modern Putting Stroke and the Modern Putter

7.1 A Short History of Green Speeds

The relationship of today's fast greens to the very recent development of the modern putting stroke and the modern putter (lower loft, more weight)

**Putting was more like a "chip and run" shot for hundreds of years;
the modern pendulum putting stroke is a <u>very</u> recent development**

Early greens were very slow, with grass higher than today's fairways. Putting was more like chipping. Early putters had 10° of loft or more (sometimes a lot more than 10°). Many golfers would carry two putters; one with about 10° of loft for putting very near the hole itself; and a second, with 12° to 15° of loft, for "putting" from distances of 30, 50 or even 80 yards from the hole. Putters through 1960 had more than 7° of loft because of the very slow/rough greens.

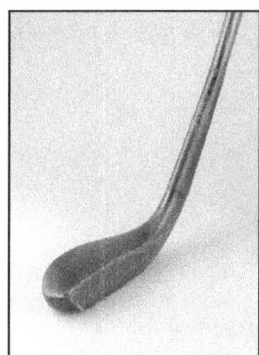

| Scythe cut greens at St Andrews in the 1800's | Old Tom Morris putting | 10° to 15° loft |

Greens in the 19th century were cut with scythes to a height that would be similar to light rough, or fairways and tees today. They would have stimped at about 3.5, maybe less. The invention of the lawn mower in 1830 by Edwin Budding in England enabled important changes in golf course maintenance.

By the early 20th century, putting greens were being mowed to 3/8" in height. By the end of the roaring 20's, green mowers had developed into specialized equipment, and mowing heights were a little tighter. Putter loft began to decrease; but the Calamity Jane putter used by Bobby Jones to win his Grand Slam in 1930 still had 8.5 degrees of loft.

The 1830 Budding lawnmower

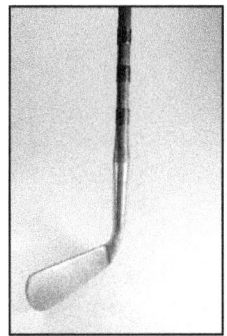
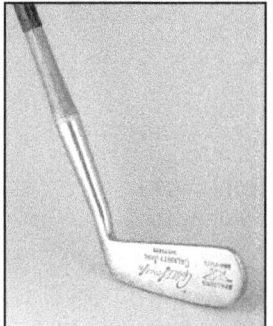

Bobby Jones's 1930's Calamity Jane putter had 8.5° loft

The idea of green speed did not even exist until the mid 1900's. Early greens were VERY slow.

97

7.2 The Evolution of the Modern Putting Stroke

The modern putting stroke did not exist prior to 1978. Old Tom Morris, Young Tom Morris, Bobby Jones, Gene Sarazen, Ben Hogan, Bobby Locke, Billy Casper and Arnold Palmer all putted on very slow greens with a (very long) POP putting stroke that was dominated by wrist cock and release.

1930
Cock & Release
Pop Stroke

Old Pop Stroke

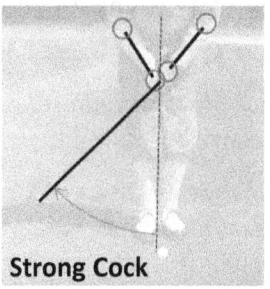 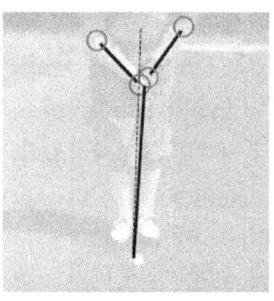 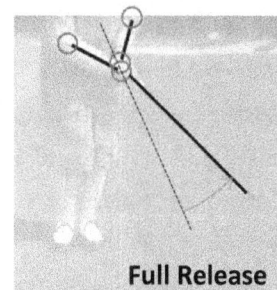

Strong Cock — **Full Release**

POP Stroke was dominated by wrist cock and release

The modern putting stroke with minimized wrist cock and release developed on the very fast greens we putt on today. Note how minimized wrist cock and release is in both the (very short) strokes below.

2020
Minimal Cock & Release
Modern Pendulum
Putting Stroke

Modern Stroke

Minimal Cock

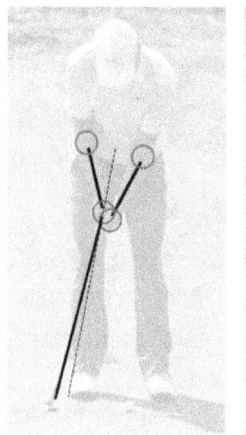 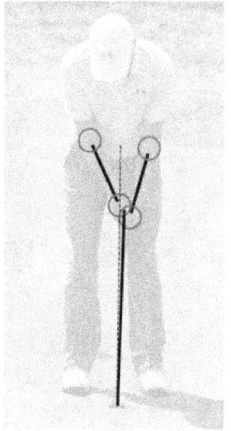 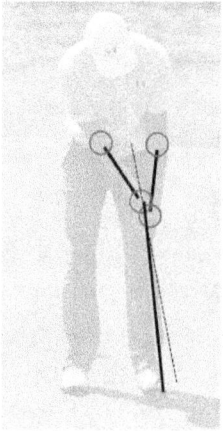

The Modern Pendulum Putting Stroke minimizes wrist cock and release

2020
Minimal Cock & Release
Modern POP
Putting Stroke

Brandt Snedecker

Modern Pop

Minimal Cock
(but a little more)

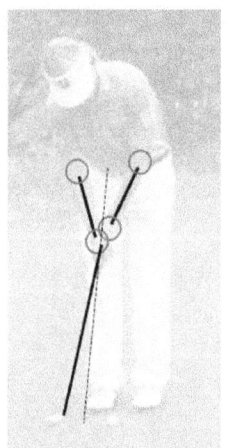 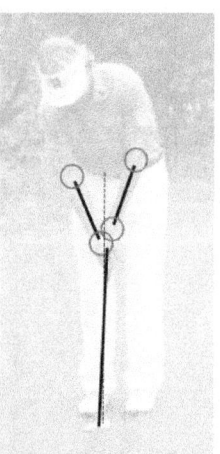 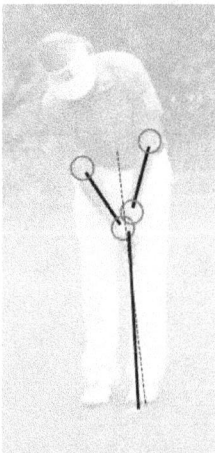

The Modern POP Putting Stroke not at all similar to the old POP stroke.
Wrist cock and release is very very limited; but more than most tour players today.

The Brandt Snedecker "POP" stroke is a modern pendulum putting stroke. His difference from other modern pendulum putting strokes is a faster tempo and a full release; compared to the dead hands modern stroke. Compared to Bobby Jones, Brandt Snedecker has almost no release at all.

7.3 The Evolution of the Modern Putter

The loft of putters decreased with lower green mow heights and higher stimps. The loft of putters through the years is far higher than most golfers would guess today.

Pre 1840
10° to 15° loft

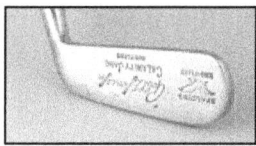
1930 Calamity Jane
8.5° loft

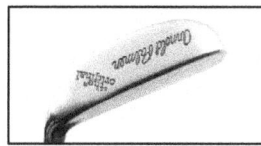
1960 Arnold Palmer
Original 7° loft

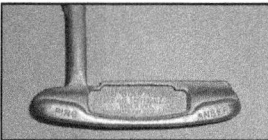
1969 Scottsdale Ping
Anser 6° loft

In 1978 the USGA used the stimpmeter to measure 581 courses nationwide to benchmark the speed of American greens. The stimps in 1978 were far lower than most golfers would guess today. Only Oakland Hills and Oakmont had stimps over 8. The average stimp on tour in 1978 was about 7. These were America's finest courses; the averages courses and "muni's" were more typically in the range of 4 to 6 feet.

1978 Green Speeds in the USA

Course	Stimp
Harbor Town	5'-1"
Congressional	6'-4"
Merion	6'-4"
San Francisco Golf Club	6'-5"
Pinehurst No. 2	6'-10"
Pebble Beach	7'-2"
Shinnecock Hills	7'-2"
Pine Valley	7'-4"
Winged Foot	7'-5"
Medinah	7'-8"
Cypress Point	7'-8"
Augusta National	7'-11"
Oakland Hills	8'-5"
Oakmont	9'-8"

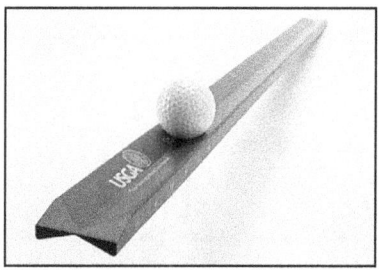

The green speeds at America's finest courses were not very fast in 1978.
The average stimp on tour was about **7** in 1978; really, only **7**.

Putter Loft and Weight Evolution

The trend toward lower loft and higher weight is evident in the table below:

Year		Putter Loft	Mowing Height	Putter Weight	Est Stimp
Pre 1850	(with scythes)	10° +	0.500"		3.5'
1920	(with mowers)	9-10°	0.375"		5.0'
1930	Bobby Jones	8.5°	0.300"	310 g	5.5'
1950	Ben Hogan	7°	0.250"	290 g	6.0'
1960	Arnold Palmer	7°	0.225"	307 g	6.5'
1970	Scottsdale Ping Anser	6°	0.210"	310 g	7.0'
1980	Carbite Z4	4°	0.190"	325 g	8.0'
1990	Rossie II	4°	0.150"	325 g	9.0'
2000	Newport 2; 2 Ball	4°	0.140"	330 g	9.5'
2010	Spider Tour	3.5°	0.130"	350 g	10.0'
2007	Rife 2 Bar	2°	0.125"	350 g	10.5
2015	Cure	1.5°	0.120"	400 - 500 g	11.0'
		trend lower		trend higher	

The Rife 2-bar putter was introduced in 2007. Rife's marketing claimed that the great roll their putter produced was caused by their patented groove design. The 2° loft (much lower loft the competition) was much more important to their success than their grooves. Today's greens argue for lower loft and higher weight putters. The tale will be told in putting launch monitor data; but the trends are clear.

The Nested Ball Myth

The ball is no longer sitting in a deep depression in the grass on a putting green today; and the loft of putters has been steadily decreasing with mow height on the greens.

A famous putter designer in his current putter fitting guide states that his "Putter Studio research shows that a ball pushes down slightly into the grass on a green, and that 3.5° loft is needed to lift the ball up and on to the surface for a smooth roll." This is a verbatim quote. He is not alone. This statement could have been made by any of hundreds of top teachers. It is no longer true. It's a myth.

The putting stroke employed by ALL players as recently as between 1930 and 1970 was a hands/wrists pop stroke and utilized a relatively lofted putter to get the ball up (out of its nest) and rolling.

The evolution from the old pop stroke to the modern pendulum putting stroke was complete by about 2000. The evolution from the hands/wrist-based pop stroke into the modern pendulum putting stroke was a direct result of the decreased green mowing heights and the higher resulting stimps. Greenskeepers have changed the game.

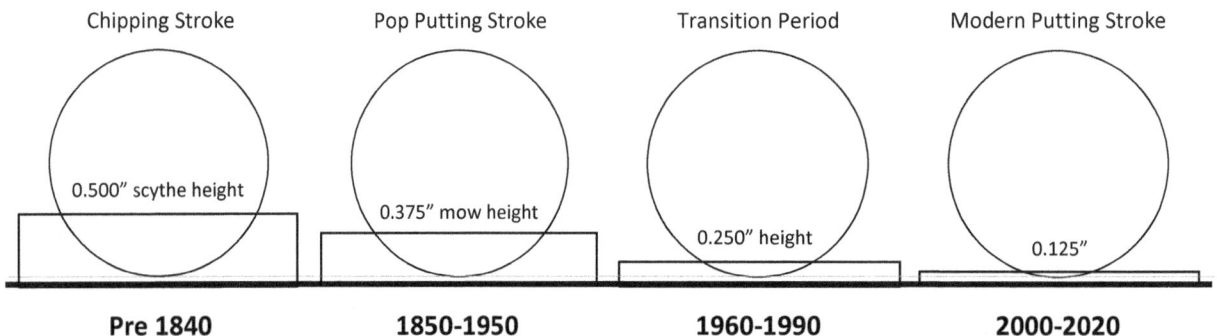

Representative Players Stroke Types Through the Green Speeds Ages:

Old Tom Morris
    Bobby Jones
        Gene Sarazen, Byron Nelson, Ben Hogan
            Arnold Palmer
                Tom Watson
                    Ben Crenshaw
                        Tiger Woods
                        & all tour players today

The Evolution of the Modern Putting Stroke (and the Modern Putter)

The evolution into the modern pendulum putting stroke is complete; but many of its characteristics (like pendulum tempo and biomechanical posture issues) have not yet been explored. The evolution of the modern putter is not complete. Most putters today have more loft than required.

The importance of moment of inertia (MOI) has been obscured by the marketing efforts of the big four putter companies (Odyssey, Ping, TaylorMade, Scotty Cameron).

The recent discovery of weight optimization putter fitting has not been widely utilized, mainly because the putters available from the big four offer such a limited weight range, if they are weight adjustable at all.

Section 8
Putting Myths and Misunderstandings

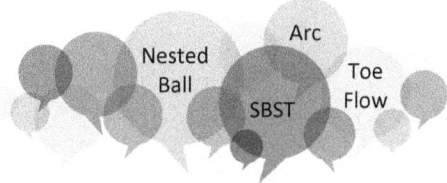

Putting Myths and Misunderstandings
8.1 Putting Myths and Misunderstandings

The Toe-Flow Myth

Confusion is created by the term "toe-flow", which has fostered the impression that a heel shafted, toe hang putter promotes rotation when it actually does the opposite. It resists rotation. Some even saying that a "toe flow" putter is easier to swing on an arc than a face balanced putter. It *is* easier to feel the rotation with a toe hang (toe flow) putter; but this is precisely because it is resisting the rotation.

Tiger Woods has a relatively strong arc in his putting stroke and likes to "feel" the toe releasing in his stroke. To "feel" this toe release, actually a very small, very very subtle, hands/wrists/forearms release in his stroke Tiger uses a mid-hang putter. He does not use a mid-hang putter to promote rotation, or because it is easier to rotate. He uses it because it resists rotation and he can "feel" the rotation/release of the toe in his stroke.

The Increased MOI of Toe-Hang Putters

Increased toe hang increases the putter's resistance to twisting on the axis of the putter shaft. In terms of moment of inertia (MOI), there is no change in the MOI of the putter head itself as measured on the vertical axis through the center of mass; but an important MOI change in the axis of the shaft.

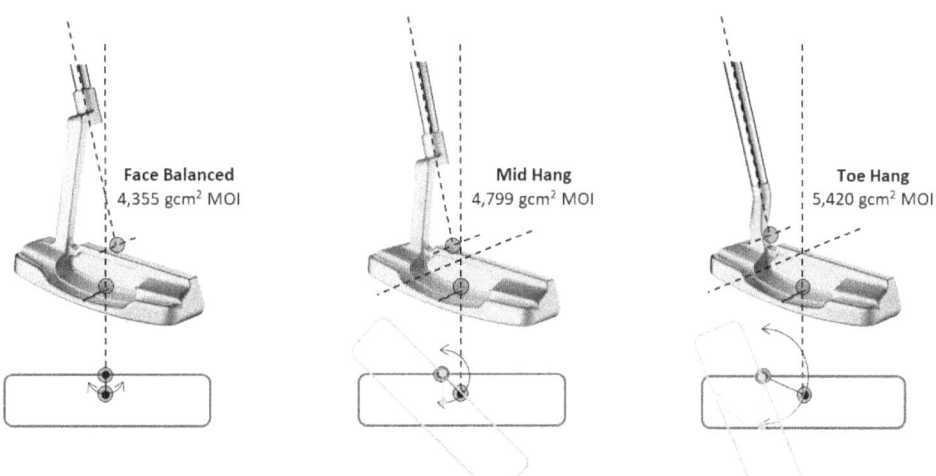

Toe weight, toe-hang, toe-flow = resistance to rotation. Period.

When is a Toe-Hang the Right Answer?

The "rule-of-thumb" recommending a toe-hang putter for a strong arc is generally only correct for a player with a strong arc produced by excessive hands/wrists/forearms actions. In this case, as an older club fitter can tell you, adding toe-hang to the putter will quiet the hands and eliminate a little of the unwanted extra hand/wrist arc. Many old-time fitters knew to add toe hang to reduce unwanted rotation in the stroke.

If the strong arc is a neutral arc related to posture and putter lie angle with no hand/wrist/forearm action, there will not necessarily be a benefit from the "feeling" of a toe-hang putter. It is possible for the increased MOI of a toe-hang putter to reduce face angle variation in the putting stroke; but simply increasing the MOI of the putter can do the same thing without introducing a new feeling to the stroke.

If a putting stroke is square to the arc, with no rotation in addition to the neutral arc rotation, then the only reason to add toe hang or toe flow in an arced putting stroke is feel.

Toe-Hang Physics

The physics is the same with a putter or with the driver. Adding weight to the toe increases resistance to toe rotation; and reducing toe weighting promotes rotation of the toe.

More weight to the toe promotes a fade by slowing the toe (resisting rotation)

Less weight to the toe promotes a draw by speeding up the toe (promoting rotation)

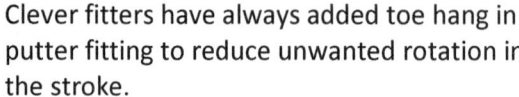

Clever fitters have always added toe hang in putter fitting to reduce unwanted rotation in the stroke.

Adding toe weight in a driver similarly reduces the rotation of head, in the case of the driver, promoting a fade.

Toe weight = slow toe rotation
In both drivers and putters.

The BIG Myth: High MOI Putters are only for High Handicap Golfers

It has been suggested that high MOI putters are for players with inconsistent strokes; i.e. high MOI putters are only for high handicap players. This is like arguing that only high handicap players will benefit from a more stable putter. Tour players tend to want workable blades; but don't necessarily want a workable putter. We are not trying to impart spin or vary the launch angle of putts. All players, regardless of skill level tend to want a more stable, higher MOI, putter; and the use of higher MOI putters on tour has been growing.

The Nested Ball Myth

The ball is no longer "nested" in the grass on the putting green. The modern putter evolved from very light wooden shafted wooden head putters with 10 or more degrees of loft (sometimes a lot more than 10 degrees). Many golfers would carry two "putters"; one with about 10 degrees of loft for putting very near the hole itself; and a second, with 12 to 15 degrees of loft, for putting from distances of 30, 50 or even 80 yards from the hole.

Greens in the 19th century were cut with scythes to a height that would be similar to our fairways and tees today. The invention of the lawn mower in 1830 by Edwin Budding in England enabled important changes in golf course maintenance.

The Stimpmeter, invented by Eddie Stimpson, was used by the USGA in 1977 to measure 581 courses nationwide to benchmark the speed of American greens. Only Oakland Hills and Oakmont had stimps over 8. The average stimp on tour in 1977 was about 7.

But the ball is no longer sitting in a deep depression in the grass on a putting green today. The growing use of launch monitor metrics has confirmed that less loft, higher weight and higher MOI are more consistent on today's faster greens.

The Myth of the Straight Back Straight Through (SBST) Putting Stroke

Almost no one has a SBST putting stroke. But many putter fitting professionals still ask if you have one. Claiming that there are two putting stroke types:

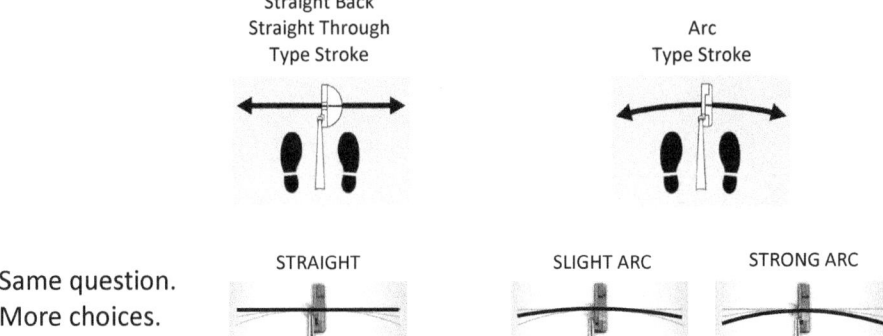

Same question. More choices.

The arc of the stroke is determined by the plane of the putting stroke; and the plane of putting stroke is strongly determined by putting posture. The myth of the SBST stroke has dominated the putting arc discussion. It is extremely rare. Almost everyone has a putting stroke arc.

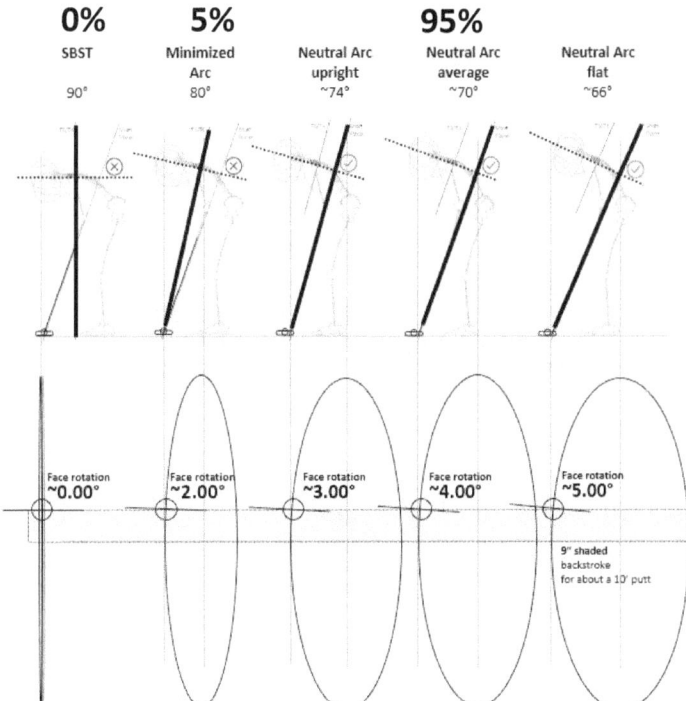

The correct question is not: Do you have an arc putting stroke? The correct questions to ask are: 1) What is your neutral arc? And 2) How much face rotation should you have to stay square to your arc? Your neutral arc is the arc created by your putting posture and the plane of your putting stroke with no excess hand and wrist action. You can know a lot about it; but it will not tell you the type of putter you should use.

The Myth of Putter Types for Putting Stroke Arc Types

A major golf instruction and club fitting company (left below) makes some interesting suggestions. A major putter manufacturer (right below) suggests face-balanced for a straight stroke, mid-hang for a slight arc, or full toe hang for a strong arc.

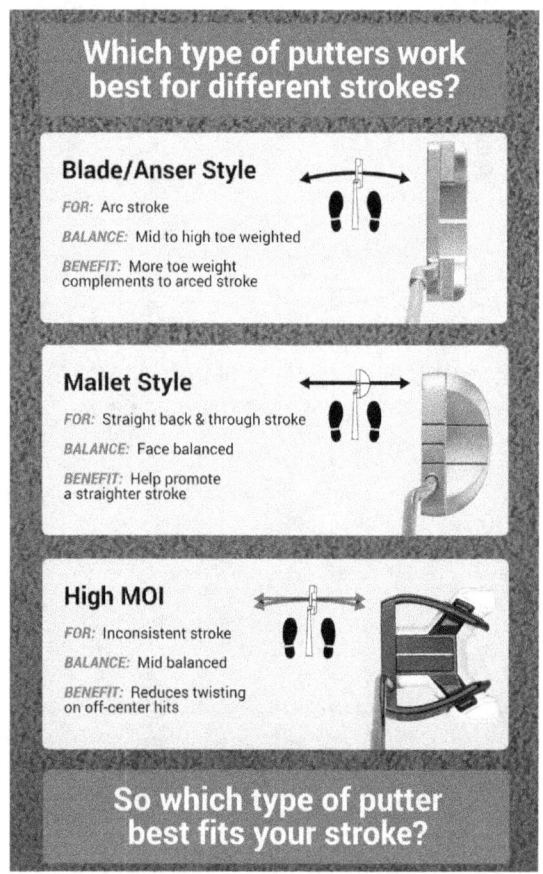

"Pick a Putter based on your Stroke Style"
golftec.com 2015

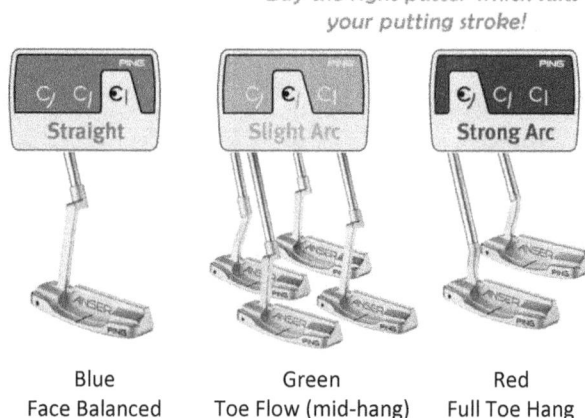

Blue — Face Balanced
Green — Toe Flow (mid-hang)
Red — Full Toe Hang

The suggestions left here are bizarre, inconsistent and incomplete. They reinforce the myth of mid to high toe weighted putters for an arc stroke. They suggest that mallets are all face balanced when they actually come in the full range of toe-hang. They recommend High MOI for inconsistent stroke, suggesting that high MOI putters are mid-balanced (mid-hang). They ignore the fact that all putter styles can be high MOI. They ignore the fact that high MOI mallets can be, in fact often are, face balanced.

Taken together, the sum of all these suggestions, is conflicted and confused. **<u>Countless examples of tour players in exactly the opposite putter from the suggestions here can be recited</u>**. Almost none of the recommendations here are based in the science of putting. This is little more than marketing gibberish.

The choice of putter type relates to personal feel, not arc type

Jack Nicklaus used a full toe-hang Low Wizard 600 for his straight back straight through stroke. The exact opposite of the marketing guys' recommendation.

Tiger Woods uses a toe-hang putter to feel the release of the putter. Steve Stricker and Loren Roberts use a full toe hang putter in a putting stroke that has much less arc, for a similar reason. Steve and Loren are known as "straight back straight through" putters because of the reduced (but still present) arc in their putting stroke. Steve and Loren do not want rotation of the toe. The toe hang putter promotes stability and allows them to feel any unwanted rotation.

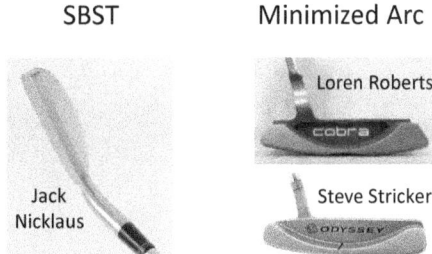

Face balanced putters are used by a large number of tour players with varying arcs in their strokes. The choice of face balanced vs mid-hang or full toe hang putters does not appear to relate to arc at all, but instead to feel and personal preference for alignment and the feel of the putter. Very personal and not related to arc.

Doubling Down on the Toe-Flow Myth

In spite of the fact that toe-hang putters have a higher shaft centered MOI and actually RESIST twisting (rotation), a number of teachers claim that toe-flow putters rotate more freely. We have recently seen a nationally known teacher claiming that "replacing a face-balanced putter with a toe-hang putter" will result in increased face rotation. This is an example of compounding a myth into to an absurdity.

First, Believing the myth that toe-hang putters rotate more freely. Then, assuming that test results would confirm more face rotation with a toe-hang putter. Actual tests have consistently confirmed that increased MOI reduces face rotation.

The examples of one-plane minimized arc putters, like Steve Stricker and Lauren Roberts, who use toe-hang putters, pointedly disprove this careless hypothesis. Players like these have reduced their face rotation so much that they are sometimes called straight-back straight-through (SBST) putters.

Why would anyone use a toe-hang putter if it actually increased face rotation?

Section 9
Putting Launch Monitor Metrics

Putting Launch Monitor Metrics

9.1 Putting Launch Monitor Metrics

Putting Launch Monitor Metrics

The growing use of putter launch monitors will dramatically change putting. Putting launch monitors differ in their technology and data available. Almost all launch monitors provide putter speed information. Not all launch monitors provide backstroke length or ball speed information.

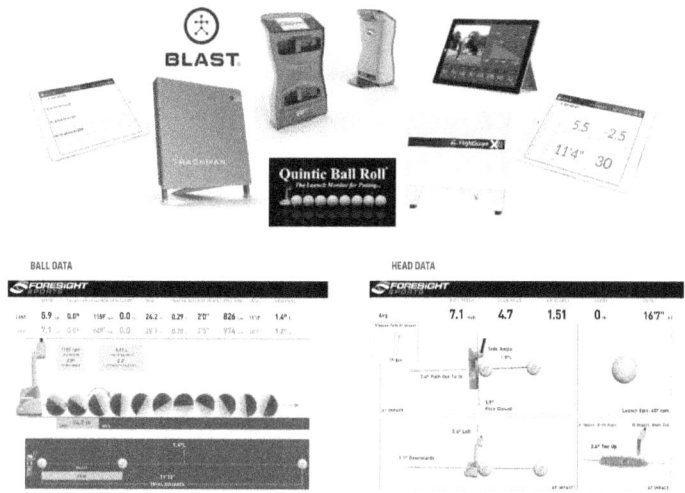

Ball data includes ball speed, launch angle, carry distance, skid distance, roll distance, side pin, top/back spin, launch direction, etc. Putter data includes face angle at impact, putter speed, backstroke length, rotation in backstroke, rotation at impact, shaft angle variation, attack angle variation, lie angle variation, club twist at impact, etc. The most important metrics are tempo and face angle at impact, and almost all launch monitors provide this data. All launch monitors provide putter speed data (see table below center column); but not all launch monitors provide ball speed and backstroke length (see table below right and left columns).

	Ball Speed	Impact Ratio	Club Speed	Calc Pmph/BS"	Backstroke Length
Blast Motion Golf	X	X	✓	✓	✓
Foresight GC Quad	✓	✓	✓	X	X
Quintic	✓	✓	✓	X	X
SAM PuttLab	X	X	✓	✓	✓
SAM PuttLab w/ Ball Tracker	✓	✓	✓	✓	✓
Trackman 4	✓	✓	✓	✓⊗	✓⊗

Finding your personal impact ratio and Pmph/BS" is not as simple today as it will be in the future. Today you have to use multiple launch monitors unless you find a SAM PuttLab with Ball Tracker. The backstroke length information from Trackman 4 is displayed rounded to even inches, so it is not very useful in calculating putter mph per backstroke inch (Pmph/BS"). Blast motion is an inexpensive launch monitor that provides useful backstroke length and putter speed information. Blast has no ball data; but can be combined with the widely available launch monitors to get impact ratio data.

The Weight Optimization Fitting Protocol (in Section 6) uses face angle variation at impact data. While the idea of face angle at impact is intuitively simple. The range of face angle at impact variation for players of differing skill levels is not well known. Everyone knows that tour players are very good putters. How good are they, at controlling face angle at impact? Quintic provides the following information regarding face angle variation:

	Face Angle Variation RANGE
Elite Tour Player	0.30° to 0.50°
Average Tour Player	0.50° to 1.00°
Low handicap Amateur (poor tour players)	1.00° to 2.00°
High Handicap Amateur	over 2.00°

This is in general agreement with the Trackman suggestion of 1.00° as a face angle variation benchmark.

Note that zero variation is not a reasonable, attainable goal. The best of the best still have 0.35° to 0.50° face angle variation at impact. Weight Optimization Fitting can improve everyone, at all skill levels. Testing with tour players have yielded a few results below 0.35° face angle variation range; none yet below 0.30° variation. But expectations are that many will improve beyond this.

Quintic provides the following information regarding putting launch monitor data:

Quintic Ball Roll Putting Data Ranges for Elite, Good, Average and Poor Players	Elite Tour Player	Scratch Player to Average Tour Player	Low Handicap Amateur	High Handicap Amateur
	Blue	Green	Amber	Red
	Elite	**Good**	**Average**	**Poor**
FACE				
Face Angle Variation RANGE	< 0.50°	0.50° to 1.00°	1.00° to 2.00°	> 2.00°
Face Rotation (degr/sec) Range	< 10°	10° to 20°	20° to 40°	> 40°
TWIST				
Club Head Twist Variation RANGE	< 0.10°	0.10° to 0.30°	0.30° to 0.75°	> 0.75°
SPEED				
Impact Ball Speed Variation RANGE	< 0.35 mph	0.35 to 0.50 mph	0.50 to 1.0 mph	> 1.00 mph
Impact Club Speed Variation Range	< 0.2 mph	0.2 to 0.35 mph	0.35 to 0.55 mph	> 0.55 mph
Impact Ratio Range	< 0.02	0.02 - 0.04	0.04 - 0.08	> 0.08
SPIN				
Zero Skid Variation (inches) RANGE	< 2.0	2.0 - 7.0	7.0 - 15.0	> 15.0
Initial Ball Roll (rpm) Range	< 15	15 - 40	40 - 75	> 75
Forward Roll (inches) Range	0	0.01 - 1.00	1.00 - 2.00	> 2.00
Side Spin (rpm) Range	< 10	10 - 20	20 - 40	> 40
LAUNCH				
Launch Angle Variation RANGE	< 1.00°	1.00° to 1.75°	1.75° to 2.50°	> 2.50°
Shaft Angle Variation Range	< 0.50°	0.50° to 1.00°	1.00° to 2.00°	> 2.00°
Attack Angle Variation Range	< 0.50°	0.50° to 1.25°	1.25° to 2.50°	> 2.50°
Lie Angle Variation Range	< 0.50°	0.50° to 1.00°	1.00° to 2.00°	> 2.00°

Blue	Elite / Best of the Best - No need to improve - just keep your results in the blue
Green	Good / Scratch to Avg Tour Player - good putter; work on key values: Face, Twist, Speed, Spin & Launch
Amber	Average / low handicap amateur - still room for improvement
Red	Poor / high handicap amateur - lots of problems putting

Focus on your personal pendulum putting tempo and on face angle variation in weight optimization fitting with a high MOI putter and ALL of your other metrics will improve together.

Shown in this table are the actual putting percentages on tour: make, two putt and three putt or more. Also shown is the width of one degree at the putt length. You can see one degree on the green.

PGA Tour Make Percentages

Putt Length in Feet	One Putt	Two Putt	Three Putt or more	Inches per one degree at Putt Length	
1	100%	0.0%	0.0%	0.21	
2	99%	0.6%	0.1%	0.42	
3	95%	5%	0.2%	0.63	
4	86%	14%	0.2%	0.84	
5	75%	25%	0.3%	1.05	
6	65%	35%	0.3%	1.26	
7	56%	44%	0.4%	1.47	
8	49%	51%	0.5%	1.68	the 1.68" ball is one degree wide at 8'
9	43%	56%	0.6%	1.89	
10	38%	61%	0.7%	2.09	
11	34%	65%	0.8%	2.30	
12	30%	69%	0.9%	2.51	
13	27%	72%	1%	2.72	
14	25%	74%	1%	2.93	
15	22%	77%	1%	3.14	
16	20%	78%	1%	3.35	
17	19%	80%	2%	3.56	
18	17%	81%	2%	3.77	
19	16%	83%	2%	3.98	
20	14%	84%	2%	4.19	the 4.25" hole is about one degree wide at 20'
21	13%	85%	2%	4.40	
24	11%	86%	3%	5.03	
27	9%	88%	4%	5.66	
30	7%	88%	5%	6.28	
40	4%	86%	10%	8.38	
50	3%	81%	16%	10.47	
60	2%	74%	24%	12.57	one degree is just over 12" at 60'
70	2%	68%	31%	14.66	
80	1%	63%	36%	16.76	
90	1%	60%	39%	18.85	
100	1%	60%	39%	20.95	

Section 10
Miscellaneous Putting Fundamentals

Miscellaneous Putting Fundamentals
10.1 Alignment, Grip and Stance

Players of all skill levels should carefully confirm where their putter is actually aimed vs where they think it is aimed. You can check your actual aim with a laser (extremely accurate) or with a friend or teacher (easiest and accurate enough). Simple process. Find a straight putt of about 8 feet. Place the ball with no logo or line visible from the top (naked ball with no alignment aid). Take your stance and aim the putter at the the center of the hole. From behind your putter, on the extended line of the putt, your friend or teacher can simply tell you where you are actually aimed or hold the putter in the alignment you have chosen and let you walk around behind the putter and see for yourself. Many amateurs are surprised to learn how poorly they actually aim their putter. We see a range of from dead on to as much as 10 inches left or right of the hole. For an 8 foot putt! Players with poor aim have learned to push or pull the putt toward the hole in their putting stroke. The problem is that calibrating the push/pull is not easy for putts of varying length; and when they grind on a short putt they often succeed in hitting it where they are aimed (not at the hole).

The perceived line of putt (from the putting stance) is an optical illusion, distorted by parallax error and eye dominance. Aligning a line on the ball to the line of the putt can eliminate this part of the optical illusion in putting. Very few tour players putt without carefully aligning the ball to the line of the putt; and then carefully aligning the putter to the line on the ball. It takes a little time; but it can fix your aim.

Aligning the Ball to the Line of the Putt
It is relatively easy to see the intended line from behind the ball standing on the extended line of putt.

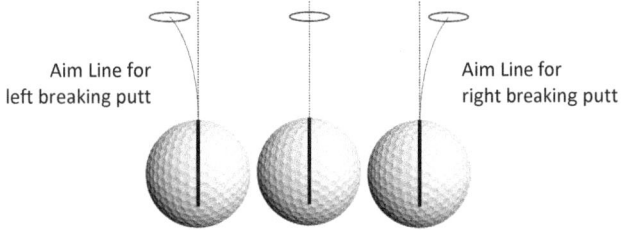

Aligning the Putter to the Ball
It is important to be very careful aligning the putter to the line on the ball. Work hard to get it right.

You need to check your putter alignment the same way you checked your aim. The easiest way is to have friend or teacher confirm your putter alignment from behind. We have seen a full range of errors here also.

First take care to align your ball; and then take care to align your putter. You can then take your putting stance over the ball and be assured that you are addressing the ball with the putter better aligned to the intended line of the putt; and take a putting stroke that requires no mid-course adjustment.

The Variety of Grips

Most tour players use the traditional right hand low grip, but many use: the left hand low grip, split grip, arm lock grip, right hand low claw grip, etc. etc.

The choice of grip is extremely personal. The reasons for choosing one grip over another include: comfort; quality of release (limited release. ease of release or even feel of release); quality of contact; arc consistency; distance control; line control; but in the end, mostly just feel.

> **NOTE:**
> The position of your hands, right hand low or left hand low, has a strong tendency to effect the alignment of the shoulders and the T3/T4 axis of rotation for the putting stroke. If your shoulders are effected by your hand position, your stance can be adjusted in response.

The Variety of Stances

A player's stance must be changed in response to changes in the putting arc caused by the grip choice. The goal is taking a stance with your biomechanical axis of rotation (T3/T4) square (as in figure A below) to the line of the putt allows the putter <u>face</u> to be <u>square</u> to the arc in order to be square to the line.

(A) The putter face is 90° to Line of Putt AT IMPACT in all three examples below, but only in (A) is the putter square to the arc at impact.

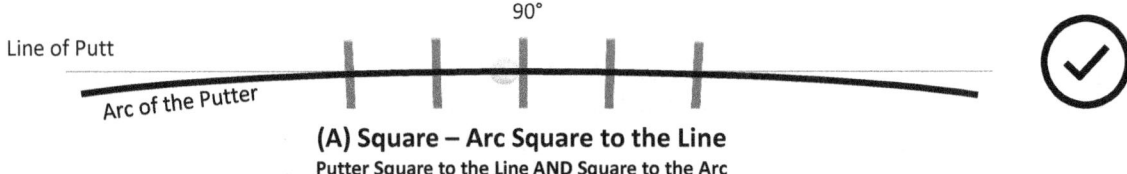

(A) Square – Arc Square to the Line
Putter Square to the Line **AND** Square to the Arc

(B) Taking a stance with your biomechanical axis of rotation (T3/T4) open (as in figure B below) requires the putter <u>face</u> to be <u>open</u> to the arc in order to be square to the line.

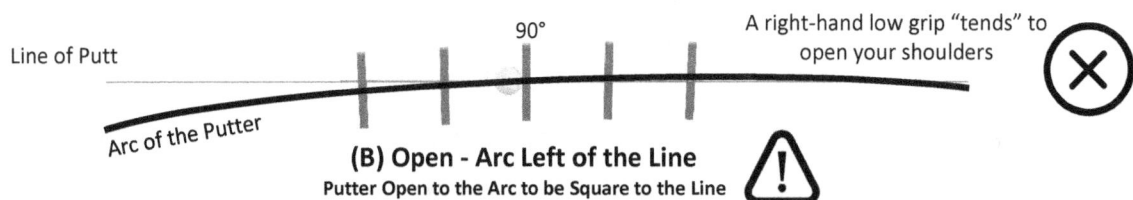

(B) Open - Arc Left of the Line
Putter Open to the Arc to be Square to the Line

(C) Taking a stance with your biomechanical axis of rotation (T3/T4) open (as in figure C below) requires the putter <u>face</u> to be <u>closed</u> to the arc in order to be square to the line.

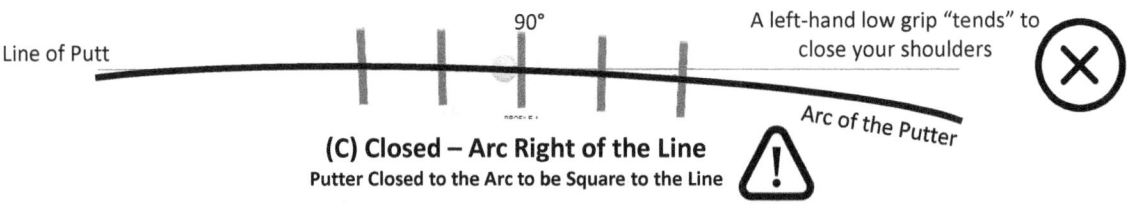

(C) Closed – Arc Right of the Line
Putter Closed to the Arc to be Square to the Line

Grip choice can impact the axis of rotation; stance must respond. The alignment of the putting arc and the line of the putt in (A) above will be easiest to maintain consistency.

Section 11
Miscellaneous Technical Information

Miscellaneous Technical Information

11.1 Excess Exit Angle from Gear Effect Spin with Rearward Center of Gravity Putters

Gear Effect Spin is increased in rearward Center of Gravity (COG) Putters and this spin causes excess exit angle for the ball; causing the putt to start off line dramatically.

The twist at impact from an off-center hit causes the putter to rotate (clockwise in the figure below). This face rotation is transferred to the ball during contact, causing the ball to rotate in the opposite direction (counterclockwise in the figure below). <u>In the air, this is hook spin</u>. Gear effect hook spin (in the air) draws back to the intended line of the shot. <u>On the green, the same hook spin causes the ball to squirt to the right</u>, like right English on a billiard ball which curves to the right. But with a putt it essentially jumps or bounces to the right because the spin is largely consumed in the initial contact with the ground.

The initial bounce to the right (on toe-hit putts), the excess exit angle from gear effect spin, appears to be on the order of 5 times greater than the twist at impact. Quintic Ball Roll Labs testing showed that a mid-mallet putter twisted less than 0.25 degrees during impact with the ball (about 0.00043 seconds) but <u>the exit angle of the ball was a whopping 1.25 degrees offline</u> (5 times the twist at impact).

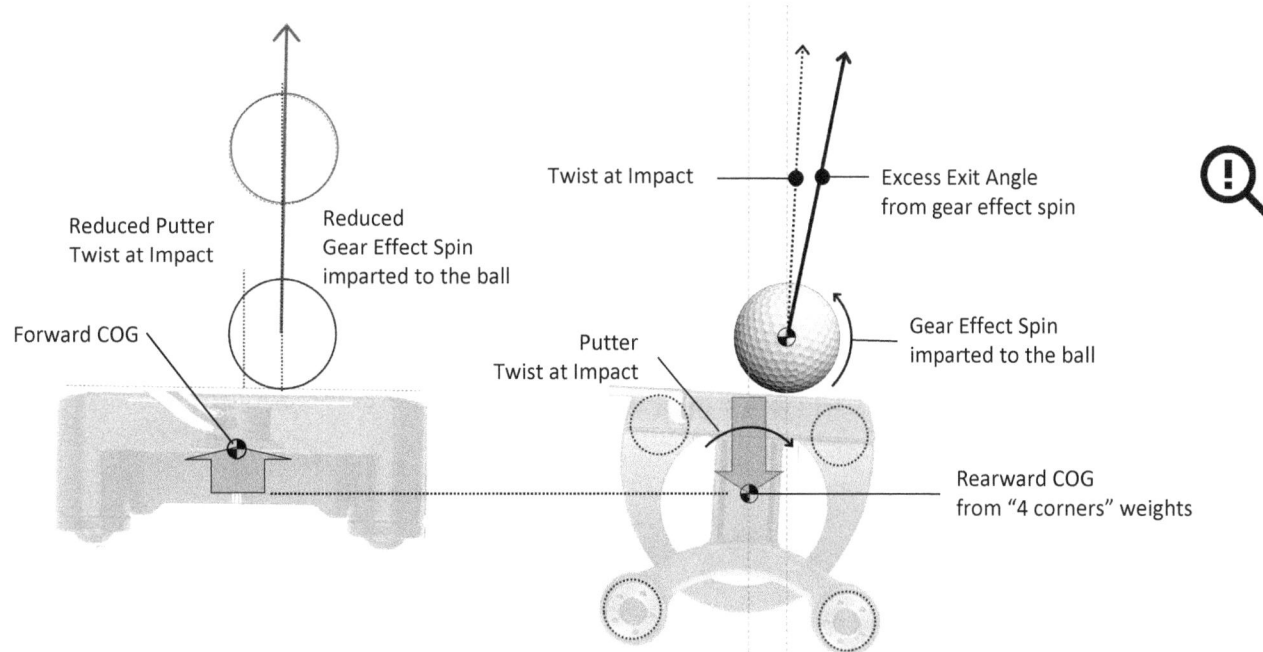

An observation about the excess exit angle gear effect spin issues with rearward COG high MOI putters:

> "More often than not the size of the head needs to get larger to make large gains in MOI. Also, designs also need to become more ring-like in appearance to really enhance the MOI while keeping the footprint of the putter small. The other disadvantage to chasing large MOI in a design can be that the CG of the putter gets deeper in the putter head. This can cause more movement in the face laterally for off-center hits. This creates **<u>larger side angle discrepancies</u>** even with large MOI values. In good designs, we look to optimize MOI while still working to keep the CG as shallow as possible in the design."
>
> ***Odyssey's Chief Putter Designer, Austie Rollinson***

Austie Rollinson calls the "excess exit angle" in the figure above right "larger side angle discrepancies"; and states that good designs "optimize MOI while still working to keep the CG as shallow as possible." Forward center of gravity (COG) is an important goal in putter design design and engineering.

Gear Effect Spin Geometry

Gear effect spin is created by the interaction between the ball and face of the putter. The lateral displacement (movement to the right in the figures below) of the contact point while the face is in contact with the ball causes the ball to spin. The diagrams below exaggerate the twist of the putter head to demonstrate the principle.

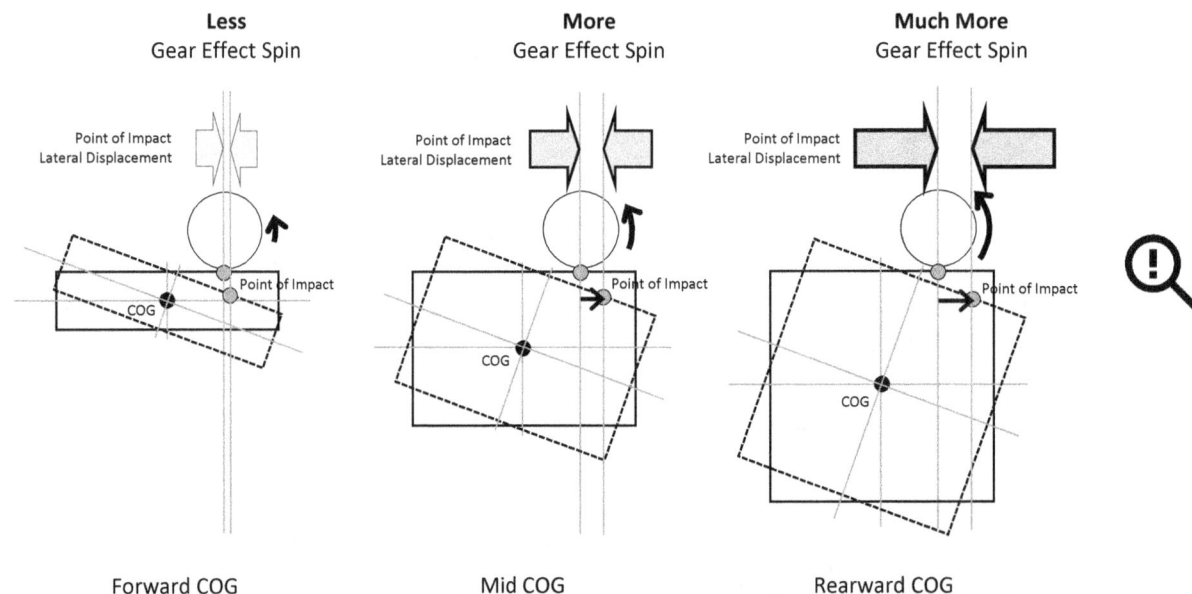

Increased Gear Effect Spin
from rearward COG
from the same amount of face rotation at impact

Gear Effect Spin is increased by rearward COG

The rearward location of the COG on large mallets and large ring or rear weighted mallets necessarily increases gear effect spin, and excess exit angle, on off-center hits.

Miscellaneous Technical Information
11.2 A Pictorial Overview of Historical Putter MOI and Radius of Gyration

MOI (gcm^2) divided by mass (g) = k^2. This is the radius of gyration^2 and it is a direct reflection of MOI weight distribution efficiency.

Putter	Mass	MOI	k^2	Width
Bullseye	304g	2,409 gcm^2	k^2 = 7.9	
8802	295g	3,381 gcm^2	k^2 = 11.5	
Newport 2	330g	4,071 gcm^2	k^2 = 12.3	
Spider 2.0	355g	5,241 gcm^2	k^2 = 14.8	
CURE TX1	380g	6,600 gcm^2	k^2 = 17.4	4.85"
CURE CX1	376g	7,200 gcm^2	k^2 = 19.1	5.00"
CURE RXi	390-410g	9,850 gcm^2	k^2 = 25.3	5.50"
Rossie II	323g	2,938 gcm^2	k^2 = 9.1	
#9/Del Mar	335g	3,484 gcm^2	k^2 = 10.4	
GoLo 5	335g	3,971 gcm^2	k^2 = 11.9	
CURE CX3	408g	8,200 gcm^2	k^2 = 20.1	5.00"
#7 Versa	345g	3,771 gcm^2	k^2 = 10.9	
EXO #7	360g	5,890 gcm^2	k^2 = 16.4	
CURE TX3	406g	7,150 gcm^2	k^2 = 17.6	4.85"
Orig 2 Ball	337g	3,553 gcm^2	k^2 = 10.5	
White Hot Pro V-Line	350g	4,164 gcm^2	k^2 = 11.9	
Spider Tour	330g	4,995 gcm^2	k^2 = 15.1	
EXO 2 Ball	360g	5,911 gcm^2	k^2 = 16.4	
CURE RX3	376-586g	7,700-14,200 gcm^2	k^2 = 20.5-24.2	5.25"
Futura X	360g	7,534 gcm^2	k^2 = 20.9	
Daddy Long Legs	395g	8,856 gcm^2	k^2 = 22.4	
CURE RX4	393-603g	8,200-14,600 gcm^2	k^2 = 20.9-24.2	5.50"
CURE RX5	405-615g	10,200-18,000 gcm^2	k^2 = 25.2-29.3	6.00"
Ping Doc17	353g	11,575 gcm^2	k^2 = 32.8	6.75"

Note that many deep shaped putters have rearward center of gravity that impacts gear effect spin. ALL of the Cure putters are designed with forward COG to reduce gear effect spin and the resulting excess exit angle problems See Section 11.1 above.

Miscellaneous Technical Information
11.3 Moment of Inertia Definitions

Scientific MOI definitions tailored to the putter industry

Inertia
A putter at rest (and square to the arc) tends to stay at rest (and square to the arc) unless acted upon by an outside force; A putter in motion (and square to the arc) tends to stay in motion (and square to the arc) unless acted upon by an outside force.

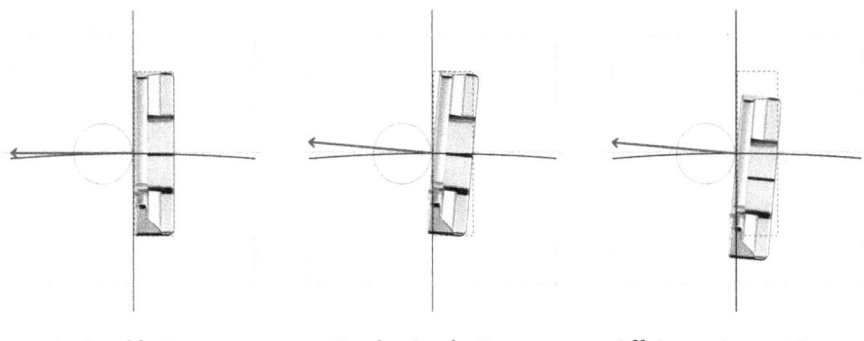

No Outside Force   Putting Stroke Forces   Off-Center Impact Forces

Putting Stroke Forces (Outside Forces) include: the variety of forces possible from the players hands, wrists, arms, shoulders (in the stroke itself);

Off-Center Impact Forces are the torque applied by an off-center hit;

and small oscillations in the shaft can add to the twist.

Moment of Inertia (MOI)
Also known as Rotational Inertia, MOI is a measure of a putter's resistance to rotation (away from square to the arc) when acted upon by an outside force.

Stability
The property of a putter in a condition of equilibrium (and square to the arc) or steady motion (and square to the arc) that resists rotational changes in its equilibrium or steady motion when acted on by an outside force.

The direct measure of a putter's actual scientific stability is its moment of inertia (MOI).

Moment of Inertia remains obscure and confusing because the marketing departments of major putter companies have dissembled and obfuscated, widely using the term "stability" as if it were independent of moment of inertia (MOI). It is not. The major putter companies continue to claim increased stability from a variety of design characteristics but insist that the actual MOI resulting from their design innovations is proprietary information. The advent of putting launch monitors will bring this dissembling to an end, eventually.

Any increase in stability will be reflected in increased MOI (as a scientific necessity).

The Physics and Geometry of MOI

The simplified formula for moment of inertia (MOI) is $I = md^2$.
  "I" is moment of inertia (gcm²)
  "m" is mass (g)
  "d" is distance (cm)

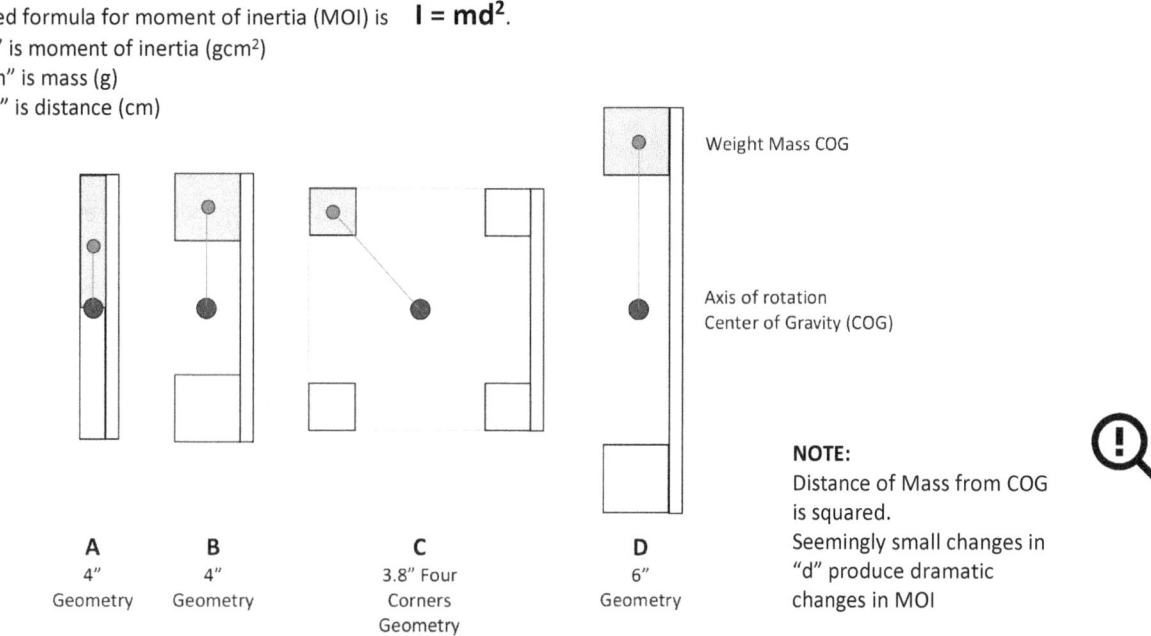

NOTE:
Distance of Mass from COG is squared.
Seemingly small changes in "d" produce dramatic changes in MOI

A	B	C	D
4" Geometry	4" Geometry	3.8" Four Corners Geometry	6" Geometry

MOI = mass x distance (from the axis of rotation) squared

In this simplified example, the putter face is assumed to be weightless and the total putter weight remains constant. The shaded mass in putter A is the toe mass in relation to the COG rotation axis of the putter (red circle). Putter A represents the old blades. Putter B represents the Ping Anser style putters. Putter C is the four corners design of many mallets. Putter D breaks the 4" geometry (like the Big Bertha broke the old driver geometry) to increase the MOI of the putter.

ALL 350 g putters

Putter A: m = **350 g** 2.50 cm d² = 6.25 cm²   MOI = 2,188 gcm²
Putter B: m = **350 g** 3.50 cm d² = 12.25 cm² MOI = 4,288 gcm²
Putter C: m = **350 g** 4.75 cm d² = 22.56 cm² MOI = 7,897 gcm²   NOTE rearward COG in putter C
Putter D: m = **350 g** 6.25 cm d² = 39.06 cm² MOI = 13,672 gcm²

This example demonstrates the importance of geometry in MOI. An increase in mass will increase the MOI of the putter linearly. An increase in distance from the center of gravity will increase the MOI exponentially. A high MOI putter resists twisting, whether from an off-center impact, or from unwanted, and unintentional, forces from hands and wrists in the stroke itself.

Miscellaneous Technical Information
11.4 The Effect of Shaft Stability on Putter Stability

Shaft Stability

A very stiff putter shaft reduces putter head oscillation and twisting during the stroke but very stiff shafts do not feel good to better players. Improved metrics but bad feel. The Stability shaft by BGT provides a very stiff shaft that feels soft and is gaining acceptance on tour. The reduction in putter twist is significant but much less than the improvement offered by high MOI. High MOI Putter testing on an optical launch monitor showed more than 50% reduction in face angle variation range from increased MOI and an additional 10% to 15% with the Stability shaft over standard steel. The Ski Pole putter shaft offered a 10% reduction but is quite heavy and often rejected by players because of its feel. The Stability shaft offers real measurable improvement but much less improvement than from MOI.

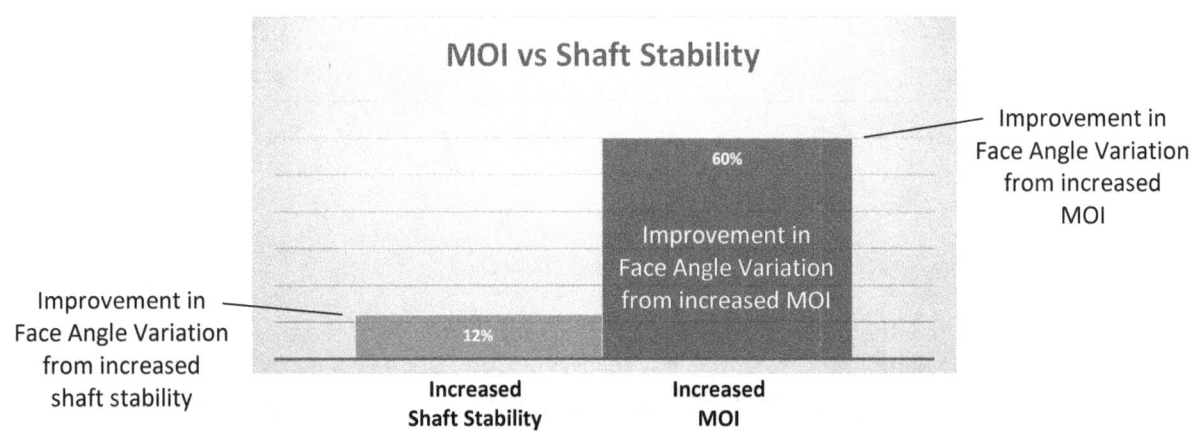

High MOI Putters offer a much larger improvement in putting stroke stability, much less twist at impact from off-center hits and have lower gear effect spin from their forward COG design.

Players can add to the high MOI advantage with a more stable shaft (the shaft stability and MOI benefits are additive); but the shaft alone does not compare to the high MOI advantage.

Miscellaneous Technical Information
11.5 The Dimple Effect

Modern golf balls have between 328 and 376 dimples; meaning that the circumference of the ball is about 32 dimples. The 32 sided polygon below is representative of a golf ball. Each side is 11.25° from its neighbor. The surface of a golf ball is faceted, not spherical. The dimples on a golf ball help us hit long and straight shots in the air; but these same dimples are a problem in putting in two ways: 1) dimples cause a rolling ball to <u>wobble</u> as it slows or stops on very fast greens; and 2) putter impact at the edge of a dimple will deflect the golf ball <u>off line</u>.

Dimple Effect Illustrated

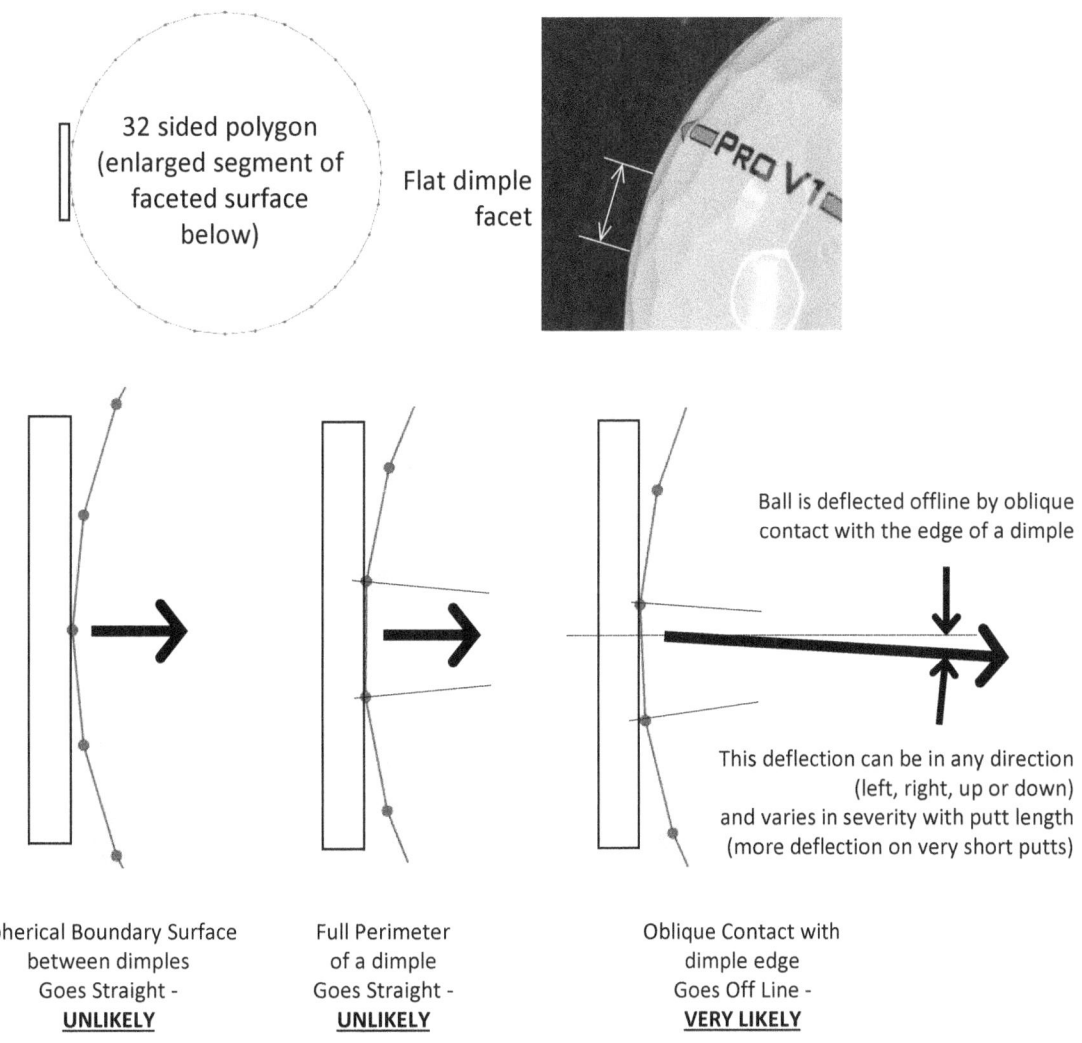

The magnitude of the deflections are not insignificant, often more than a full degree off line. The dimple effect disappears on putts longer than about 6 feet because the ball cover is compressed.

Very soft inserts can reduce dimple effect, but cannot eliminate it. An aluminum face (because it is softer) will have less dimple effect than a steel face. The USGA/R&A may eventually allow a smooth ball with no dimples to be used on greens to eliminate dimple effect.

The dimple effect is a dirty secret in the golf industry. All manufacturers know about it.
No one talks about it. We almost left it out of this book, but chose to include it to start a discussion.

Appendix A
Backstroke Length Cards

Backstroke Length Cards
Side A
Effective Stimp on Sloping Greens

stimp	6	7	8	9	10	11	12	13	14
up 6%	4.02	4.48	4.91	5.31	5.67	6.03	6.36	6.67	6.97
up 5%	4.26	4.77	5.25	5.70	6.11	6.51	6.89	7.24	7.59
up 4%	4.53	5.10	5.64	6.15	6.62	7.08	7.52	7.93	8.33
up 3%	4.83	5.47	6.09	6.67	7.23	7.76	8.28	8.76	9.24
up 2%	5.16	5.90	6.61	7.30	7.95	8.59	9.21	9.80	10.38
up 1%	5.55	6.40	7.24	8.06	8.85	9.64	10.41	11.15	11.90
level	6.00	7.00	8.00	9.00	10.00	11.00	12.00	13.00	14.00
down 1%	6.45	7.63	8.85	10.10	11.39	12.72	14.10	15.51	16.98
down 2%	7.01	8.44	9.98	11.63	13.40	15.32	17.41	19.67	22.16
down 3%	7.71	9.51	11.55	13.87	16.52	19.63	23.31	27.73	33.18
down 4%	8.63	10.99	13.86	17.44	22.01	28.13	36.74	49.79	72.10
down 5%	9.85	13.12	17.56	23.93	33.93	52.00	94.97		
down 6%	11.55	16.47	24.40	39.42	79.28				

Side B
Backstroke Length for Varying Pmph/BS"
Choose a side B for your personal card from cards on the following pages.

Note:
Two variables:

Pmph/BS" and Impact Ratio
Average is likely 0.36 Pmph/BS" and 1.75 Impact Ratio
But can vary widely!

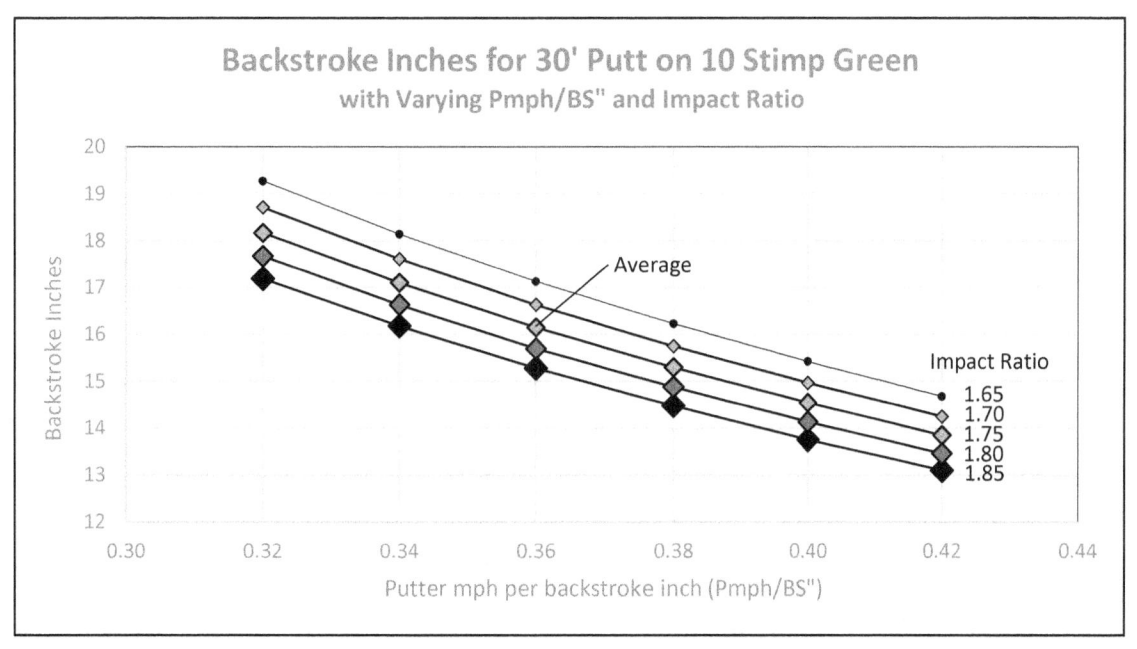

1.65 Impact Ratio

Pmph/BS″	0.32	4	5	6	7	8	9	10	11	12	13	14	15	16	17	18	19	20	25	30	40	50	1.65	Impact Ratio
	8 inches past hole																							
	3	10.5	9.4	8.6	8.0	7.5	7.0	6.7	6.4	6.1	5.8	5.6	5.4	5.3	5.1	5.0	4.8	4.7	4.2	3.8	3.3	3.0	3	
	6	14.2	12.7	11.6	10.7	10.0	9.5	9.0	8.6	8.2	7.9	7.6	7.3	7.1	6.9	6.7	6.5	6.4	5.7	5.2	4.5	4.0	6	
	9	17.1	15.3	14.0	12.9	12.1	11.4	10.8	10.3	9.9	9.5	9.1	8.8	8.6	8.3	8.1	7.9	7.7	6.8	6.2	5.4	4.8	9	
	12	19.6	17.5	16.0	14.8	13.8	13.1	12.4	11.8	11.3	10.9	10.5	10.1	9.8	9.5	9.2	9.0	8.8	7.8	7.2	6.2	5.5	12	
	15	21.8	19.5	17.8	16.5	15.4	14.5	13.8	13.1	12.6	12.1	11.6	11.2	10.9	10.6	10.3	10.0	9.7	8.7	8.0	6.9	6.2	15	
	18	23.8	21.3	19.4	18.0	16.8	15.9	15.0	14.3	13.7	13.2	12.7	12.3	11.9	11.5	11.2	10.9	10.6	9.5	8.7	7.5	6.7	18	
	21	25.6	22.9	20.9	19.4	18.1	17.1	16.2	15.4	14.8	14.2	13.7	13.2	12.8	12.4	12.1	11.8	11.5	10.2	9.4	8.1	7.2	21	
	24	27.3	24.4	22.3	20.7	19.3	18.2	17.3	16.5	15.8	15.2	14.6	14.1	13.7	13.3	12.9	12.5	12.2	10.9	10.0	8.6	7.7	24	
	27	28.9	25.9	23.6	21.9	20.5	19.3	18.3	17.5	16.7	16.1	15.5	14.9	14.5	14.0	13.6	13.3	12.9	11.6	10.6	9.2	8.2	27	
	30	30.5	27.3	24.9	23.0	21.5	20.3	19.3	18.4	17.6	16.9	16.3	15.7	15.2	14.8	14.4	14.0	13.6	12.2	11.1	9.6	8.6	30	
	40	35.1	31.4	28.7	26.5	24.8	23.4	22.2	21.2	20.3	19.5	18.8	18.1	17.5	17.0	16.5	16.1	15.7	14.0	12.8	11.1	9.9	40	
	50	39.2	35.0	32.0	29.6	27.7	26.1	24.8	23.6	22.6	21.7	20.9	20.2	19.6	19.0	18.5	18.0	17.5	15.7	14.3	12.4	11.1	50	
	60	42.9	38.3	35.0	32.4	30.3	28.6	27.1	25.8	24.7	23.8	22.9	22.1	21.4	20.8	20.2	19.7	19.2	17.1	15.7	13.6	12.1	60	
	70	46.3	41.4	37.8	35.0	32.7	30.8	29.3	27.9	26.7	25.7	24.7	23.9	23.1	22.4	21.8	21.2	20.7	18.5	16.9	14.6	13.1	70	
	80	49.4	44.2	40.4	37.4	34.9	33.0	31.3	29.8	28.5	27.4	26.4	25.5	24.7	24.0	23.3	22.7	22.1	19.8	18.0	15.6	14.0	80	
	90	52.4	46.9	42.8	39.6	37.1	34.9	33.1	31.6	30.3	29.1	28.0	27.1	26.2	25.4	24.7	24.0	23.4	21.0	19.1	16.6	14.8	90	
	100	55.2	49.4	45.1	41.7	39.0	36.8	34.9	33.3	31.9	30.6	29.5	28.5	27.6	26.8	26.0	25.3	24.7	22.1	20.2	17.5	15.6	100	

0.34	4	5	6	7	8	9	10	11	12	13	14	15	16	17	18	19	20	25	30	40	50	1.65
8 inches past hole																						
3	9.9	8.9	8.1	7.5	7.0	6.6	6.3	6.0	5.7	5.5	5.3	5.1	5.0	4.8	4.7	4.6	4.4	4.0	3.6	3.1	2.8	3
6	13.4	12.0	10.9	10.1	9.5	8.9	8.5	8.1	7.7	7.4	7.1	6.9	6.7	6.5	6.3	6.1	6.0	5.3	4.9	4.2	3.8	6
9	16.1	14.4	13.1	12.2	11.4	10.7	10.2	9.7	9.3	8.9	8.6	8.3	8.1	7.8	7.6	7.4	7.2	6.4	5.9	5.1	4.6	9
12	18.4	16.5	15.1	13.9	13.0	12.3	11.7	11.1	10.6	10.2	9.9	9.5	9.2	8.9	8.7	8.5	8.2	7.4	6.7	5.8	5.2	12
15	20.5	18.3	16.7	15.5	14.5	13.7	13.0	12.4	11.8	11.4	11.0	10.6	10.3	9.9	9.7	9.4	9.2	8.2	7.5	6.5	5.8	15
18	22.4	20.0	18.3	16.9	15.8	14.9	14.2	13.5	12.9	12.4	12.0	11.6	11.2	10.9	10.5	10.3	10.0	9.0	8.2	7.1	6.3	18
21	24.1	21.6	19.7	18.2	17.0	16.1	15.2	14.5	13.9	13.4	12.9	12.4	12.1	11.7	11.4	11.1	10.8	9.6	8.8	7.6	6.8	21
24	25.7	23.0	21.0	19.4	18.2	17.1	16.3	15.5	14.9	14.3	13.7	13.3	12.9	12.5	12.1	11.8	11.5	10.3	9.4	8.1	7.3	24
27	27.2	24.4	22.2	20.6	19.3	18.2	17.2	16.4	15.7	15.1	14.6	14.1	13.6	13.2	12.8	12.5	12.2	10.9	9.9	8.6	7.7	27
30	28.7	25.7	23.4	21.7	20.3	19.1	18.1	17.3	16.6	15.9	15.3	14.8	14.3	13.9	13.5	13.2	12.8	11.5	10.5	9.1	8.1	30
40	33.0	29.5	27.0	25.0	23.4	22.0	20.9	19.9	19.1	18.3	17.7	17.1	16.5	16.0	15.6	15.2	14.8	13.2	12.1	10.4	9.3	40
50	36.9	33.0	30.1	27.9	26.1	24.6	23.3	22.2	21.3	20.4	19.7	19.0	18.4	17.9	17.4	16.9	16.5	14.7	13.5	11.7	10.4	50
60	40.3	36.1	32.9	30.5	28.5	26.9	25.5	24.3	23.3	22.4	21.6	20.8	20.2	19.6	19.0	18.5	18.0	16.1	14.7	12.8	11.4	60
70	43.5	38.9	35.5	32.9	30.8	29.0	27.5	26.3	25.1	24.2	23.3	22.5	21.8	21.1	20.5	20.0	19.5	17.4	15.9	13.8	12.3	70
80	46.5	41.6	38.0	35.2	32.9	31.0	29.4	28.1	26.9	25.8	24.9	24.0	23.3	22.6	21.9	21.3	20.8	18.6	17.0	14.7	13.2	80
90	49.3	44.1	40.3	37.3	34.9	32.9	31.2	29.7	28.5	27.4	26.4	25.5	24.7	23.9	23.2	22.6	22.1	19.7	18.0	15.6	13.9	90
100	52.0	46.5	42.4	39.3	36.7	34.6	32.9	31.3	30.0	28.8	27.8	26.8	26.0	25.2	24.5	23.8	23.2	20.8	19.0	16.4	14.7	100

0.36	4	5	6	7	8	9	10	11	12	13	14	15	16	17	18	19	20	25	30	40	50	1.65
8 inches past hole																						
3	9.4	8.4	7.6	7.1	6.6	6.2	5.9	5.6	5.4	5.2	5.0	4.8	4.7	4.5	4.4	4.3	4.2	3.7	3.4	3.0	2.6	3
6	12.6	11.3	10.3	9.5	8.9	8.4	8.0	7.6	7.3	7.0	6.8	6.5	6.3	6.1	6.0	5.8	5.6	5.1	4.6	4.0	3.6	6
9	15.2	13.6	12.4	11.5	10.8	10.1	9.6	9.2	8.8	8.4	8.1	7.9	7.6	7.4	7.2	7.0	6.8	6.1	5.6	4.8	4.3	9
12	17.4	15.6	14.2	13.2	12.3	11.6	11.0	10.5	10.1	9.7	9.3	9.0	8.7	8.4	8.2	8.0	7.8	7.0	6.4	5.5	4.9	12
15	19.4	17.3	15.8	14.6	13.7	12.9	12.2	11.7	11.2	10.7	10.3	10.0	9.7	9.4	9.1	8.9	8.7	7.7	7.1	6.1	5.5	15
18	21.1	18.9	17.3	16.0	14.9	14.1	13.4	12.7	12.2	11.7	11.3	10.9	10.6	10.3	10.0	9.7	9.5	8.5	7.7	6.7	6.0	18
21	22.8	20.4	18.6	17.2	16.1	15.2	14.4	13.7	13.1	12.6	12.2	11.8	11.4	11.0	10.7	10.4	10.2	9.1	8.3	7.2	6.4	21
24	24.3	21.7	19.8	18.4	17.2	16.2	15.4	14.6	14.0	13.5	13.0	12.5	12.1	11.8	11.5	11.1	10.9	9.7	8.9	7.7	6.9	24
27	25.7	23.0	21.0	19.4	18.2	17.2	16.3	15.5	14.9	14.3	13.8	13.3	12.9	12.5	12.1	11.8	11.5	10.3	9.4	8.1	7.3	27
30	27.1	24.2	22.1	20.5	19.2	18.1	17.1	16.3	15.6	15.0	14.5	14.0	13.5	13.1	12.8	12.4	12.1	10.8	9.9	8.6	7.7	30
40	31.2	27.9	25.5	23.6	22.1	20.8	19.7	18.8	18.0	17.3	16.7	16.1	15.6	15.1	14.7	14.3	14.0	12.5	11.4	9.9	8.8	40
50	34.8	31.1	28.4	26.3	24.6	23.2	22.0	21.0	20.1	19.3	18.6	18.0	17.4	16.9	16.4	16.0	15.6	13.9	12.7	11.0	9.8	50
60	38.1	34.1	31.1	28.8	26.9	25.4	24.1	23.0	22.0	21.1	20.4	19.7	19.0	18.5	18.0	17.5	17.0	15.2	13.9	12.0	10.8	60
70	41.1	36.8	33.6	31.1	29.1	27.4	26.0	24.8	23.7	22.8	22.0	21.2	20.6	19.9	19.4	18.9	18.4	16.4	15.0	13.0	11.6	70
80	43.9	39.3	35.9	33.2	31.1	29.3	27.8	26.5	25.4	24.4	23.5	22.7	22.0	21.3	20.7	20.2	19.6	17.6	16.0	13.9	12.4	80
90	46.6	41.7	38.0	35.2	32.9	31.1	29.5	28.1	26.9	25.8	24.9	24.1	23.3	22.6	22.0	21.4	20.8	18.6	17.0	14.7	13.2	90
100	49.1	43.9	40.1	37.1	34.7	32.7	31.0	29.6	28.3	27.2	26.2	25.3	24.5	23.8	23.1	22.5	21.9	19.6	17.9	15.5	13.9	100

1.65 Impact Ratio

Pmph/BS"

0.38	4	5	6	7	8	9	10	11	12	13	14	15	16	17	18	19	20	25	30	40	50	1.65
8 inches past hole																						
3	8.9	7.9	7.2	6.7	6.3	5.9	5.6	5.4	5.1	4.9	4.7	4.6	4.4	4.3	4.2	4.1	4.0	3.5	3.2	2.8	2.5	3
6	12.0	10.7	9.8	9.0	8.5	8.0	7.6	7.2	6.9	6.6	6.4	6.2	6.0	5.8	5.6	5.5	5.4	4.8	4.4	3.8	3.4	6
9	14.4	12.9	11.8	10.9	10.2	9.6	9.1	8.7	8.3	8.0	7.7	7.4	7.2	7.0	6.8	6.6	6.4	5.8	5.3	4.6	4.1	9
12	16.5	14.8	13.5	12.5	11.7	11.0	10.4	9.9	9.5	9.1	8.8	8.5	8.2	8.0	7.8	7.6	7.4	6.6	6.0	5.2	4.7	12
15	18.3	16.4	15.0	13.9	13.0	12.2	11.6	11.1	10.6	10.2	9.8	9.5	9.2	8.9	8.6	8.4	8.2	7.3	6.7	5.8	5.2	15
18	20.0	17.9	16.3	15.1	14.2	13.3	12.7	12.1	11.6	11.1	10.7	10.3	10.0	9.7	9.4	9.2	9.0	8.0	7.3	6.3	5.7	18
21	21.6	19.3	17.6	16.3	15.3	14.4	13.6	13.0	12.5	12.0	11.5	11.1	10.8	10.5	10.2	9.9	9.6	8.6	7.9	6.8	6.1	21
24	23.0	20.6	18.8	17.4	16.3	15.3	14.6	13.9	13.3	12.8	12.3	11.9	11.5	11.2	10.8	10.6	10.3	9.2	8.4	7.3	6.5	24
27	24.4	21.8	19.9	18.4	17.2	16.3	15.4	14.7	14.1	13.5	13.0	12.6	12.2	11.8	11.5	11.2	10.9	9.8	8.9	7.7	6.9	27
30	25.7	23.0	21.0	19.4	18.1	17.1	16.2	15.5	14.8	14.2	13.7	13.3	12.8	12.4	12.1	11.8	11.5	10.3	9.4	8.1	7.3	30
40	29.6	26.4	24.1	22.3	20.9	19.7	18.7	17.8	17.1	16.4	15.8	15.3	14.8	14.3	13.9	13.6	13.2	11.8	10.8	9.3	8.4	40
50	33.0	29.5	26.9	24.9	23.3	22.0	20.9	19.9	19.0	18.3	17.6	17.0	16.5	16.0	15.5	15.1	14.8	13.2	12.0	10.4	9.3	50
60	36.1	32.3	29.5	27.3	25.5	24.1	22.8	21.8	20.8	20.0	19.3	18.6	18.0	17.5	17.0	16.6	16.1	14.4	13.2	11.4	10.2	60
70	39.0	34.8	31.8	29.4	27.5	26.0	24.6	23.5	22.5	21.6	20.8	20.1	19.5	18.9	18.4	17.9	17.4	15.6	14.2	12.3	11.0	70
80	41.6	37.2	34.0	31.5	29.4	27.7	26.3	25.1	24.0	23.1	22.2	21.5	20.8	20.2	19.6	19.1	18.6	16.6	15.2	13.2	11.8	80
90	44.1	39.5	36.0	33.4	31.2	29.4	27.9	26.6	25.5	24.5	23.6	22.8	22.1	21.4	20.8	20.2	19.7	17.7	16.1	14.0	12.5	90
100	46.5	41.6	38.0	35.1	32.9	31.0	29.4	28.0	26.8	25.8	24.9	24.0	23.2	22.6	21.9	21.3	20.8	18.6	17.0	14.7	13.2	100

0.40	4	5	6	7	8	9	10	11	12	13	14	15	16	17	18	19	20	25	30	40	50	1.65
8 inches past hole																						
3	8.4	7.5	6.9	6.4	6.0	5.6	5.3	5.1	4.9	4.7	4.5	4.4	4.2	4.1	4.0	3.9	3.8	3.4	3.1	2.7	2.4	3
6	11.4	10.2	9.3	8.6	8.0	7.6	7.2	6.9	6.6	6.3	6.1	5.9	5.7	5.5	5.4	5.2	5.1	4.5	4.2	3.6	3.2	6
9	13.7	12.2	11.2	10.3	9.7	9.1	8.7	8.3	7.9	7.6	7.3	7.1	6.8	6.6	6.5	6.3	6.1	5.5	5.0	4.3	3.9	9
12	15.7	14.0	12.8	11.8	11.1	10.4	9.9	9.4	9.0	8.7	8.4	8.1	7.8	7.6	7.4	7.2	7.0	6.3	5.7	5.0	4.4	12
15	17.4	15.6	14.2	13.2	12.3	11.6	11.0	10.5	10.1	9.7	9.3	9.0	8.7	8.5	8.2	8.0	7.8	7.0	6.4	5.5	4.9	15
18	19.0	17.0	15.5	14.4	13.4	12.7	12.0	11.5	11.0	10.6	10.2	9.8	9.5	9.2	9.0	8.7	8.5	7.6	6.9	6.0	5.4	18
21	20.5	18.3	16.7	15.5	14.5	13.7	13.0	12.4	11.8	11.4	11.0	10.6	10.2	9.9	9.7	9.4	9.2	8.2	7.5	6.5	5.8	21
24	21.9	19.6	17.9	16.5	15.5	14.6	13.8	13.2	12.6	12.1	11.7	11.3	10.9	10.6	10.3	10.0	9.8	8.7	8.0	6.9	6.2	24
27	23.2	20.7	18.9	17.5	16.4	15.4	14.6	14.0	13.4	12.8	12.4	12.0	11.6	11.2	10.9	10.6	10.4	9.3	8.5	7.3	6.5	27
30	24.4	21.8	19.9	18.4	17.2	16.3	15.4	14.7	14.1	13.5	13.0	12.6	12.2	11.8	11.5	11.2	10.9	9.8	8.9	7.7	6.9	30
40	28.1	25.1	22.9	21.2	19.9	18.7	17.8	16.9	16.2	15.6	15.0	14.5	14.0	13.6	13.2	12.9	12.6	11.2	10.3	8.9	7.9	40
50	31.3	28.0	25.6	23.7	22.2	20.9	19.8	18.9	18.1	17.4	16.8	16.2	15.7	15.2	14.8	14.4	14.0	12.5	11.4	9.9	8.9	50
60	34.3	30.7	28.0	25.9	24.2	22.9	21.7	20.7	19.8	19.0	18.3	17.7	17.1	16.6	16.2	15.7	15.3	13.7	12.5	10.8	9.7	60
70	37.0	33.1	30.2	28.0	26.2	24.7	23.4	22.3	21.4	20.5	19.8	19.1	18.5	18.0	17.4	17.0	16.6	14.8	13.5	11.7	10.5	70
80	39.5	35.4	32.3	29.9	28.0	26.4	25.0	23.8	22.8	21.9	21.1	20.4	19.8	19.2	18.6	18.1	17.7	15.8	14.4	12.5	11.2	80
90	41.9	37.5	34.2	31.7	29.6	27.9	26.5	25.3	24.2	23.3	22.4	21.6	21.0	20.3	19.8	19.2	18.7	16.8	15.3	13.3	11.9	90
100	44.2	39.5	36.1	33.4	31.2	29.4	27.9	26.6	25.5	24.5	23.6	22.8	22.1	21.4	20.8	20.3	19.8	17.7	16.1	14.0	12.5	100

0.42	4	5	6	7	8	9	10	11	12	13	14	15	16	17	18	19	20	25	30	40	50	1.65
8 inches past hole																						
3	8.0	7.2	6.6	6.1	5.7	5.4	5.1	4.8	4.6	4.5	4.3	4.1	4.0	3.9	3.8	3.7	3.6	3.2	2.9	2.5	2.3	3
6	10.8	9.7	8.8	8.2	7.7	7.2	6.8	6.5	6.3	6.0	5.8	5.6	5.4	5.3	5.1	5.0	4.8	4.3	4.0	3.4	3.1	6
9	13.0	11.7	10.6	9.9	9.2	8.7	8.2	7.9	7.5	7.2	7.0	6.7	6.5	6.3	6.1	6.0	5.8	5.2	4.8	4.1	3.7	9
12	14.9	13.3	12.2	11.3	10.6	9.9	9.4	9.0	8.6	8.3	8.0	7.7	7.5	7.2	7.0	6.8	6.7	6.0	5.4	4.7	4.2	12
15	16.6	14.8	13.6	12.5	11.7	11.1	10.5	10.0	9.6	9.2	8.9	8.6	8.3	8.0	7.8	7.6	7.4	6.6	6.1	5.2	4.7	15
18	18.1	16.2	14.8	13.7	12.8	12.1	11.5	10.9	10.5	10.0	9.7	9.4	9.1	8.8	8.5	8.3	8.1	7.2	6.6	5.7	5.1	18
21	19.5	17.5	15.9	14.8	13.8	13.0	12.3	11.8	11.3	10.8	10.4	10.1	9.8	9.5	9.2	9.0	8.7	7.8	7.1	6.2	5.5	21
24	20.8	18.6	17.0	15.7	14.7	13.9	13.2	12.6	12.0	11.6	11.1	10.8	10.4	10.1	9.8	9.6	9.3	8.3	7.6	6.6	5.9	24
27	22.1	19.7	18.0	16.7	15.6	14.7	13.9	13.3	12.7	12.2	11.8	11.4	11.0	10.7	10.4	10.1	9.9	8.8	8.1	7.0	6.2	27
30	23.2	20.8	19.0	17.6	16.4	15.5	14.7	14.0	13.4	12.9	12.4	12.0	11.6	11.3	10.9	10.7	10.4	9.3	8.5	7.3	6.6	30
40	26.7	23.9	21.8	20.2	18.9	17.8	16.9	16.1	15.4	14.8	14.3	13.8	13.4	13.0	12.6	12.3	12.0	10.7	9.8	8.5	7.6	40
50	29.8	26.7	24.4	22.6	21.1	19.9	18.9	18.0	17.2	16.6	16.0	15.4	14.9	14.5	14.1	13.7	13.3	11.9	10.9	9.4	8.4	50
60	32.7	29.2	26.7	24.7	23.1	21.8	20.7	19.7	18.9	18.1	17.5	16.9	16.3	15.8	15.4	15.0	14.6	13.1	11.9	10.3	9.2	60
70	35.2	31.5	28.8	26.6	24.9	23.5	22.3	21.3	20.3	19.6	18.8	18.2	17.6	17.1	16.6	16.2	15.8	14.1	12.9	11.1	10.0	70
80	37.7	33.7	30.7	28.5	26.6	25.1	23.8	22.7	21.7	20.9	20.1	19.4	18.8	18.3	17.8	17.3	16.8	15.1	13.8	11.9	10.7	80
90	39.9	35.7	32.6	30.2	28.2	26.6	25.2	24.1	23.0	22.1	21.3	20.6	20.0	19.4	18.8	18.3	17.9	16.0	14.6	12.6	11.3	90
100	42.1	37.6	34.3	31.8	29.7	28.0	26.6	25.4	24.3	23.3	22.5	21.7	21.0	20.4	19.8	19.3	18.8	16.8	15.4	13.3	11.9	100

1.70 Impact Ratio

Pmph/BS"

0.32	4	5	6	7	8	9	10	11	12	13	14	15	16	17	18	19	20	25	30	40	50	1.70
	8 inches past hole																					
3	10.2	9.1	8.4	7.7	7.2	6.8	6.5	6.2	5.9	5.7	5.5	5.3	5.1	5.0	4.8	4.7	4.6	4.1	3.7	3.2	2.9	3
6	13.8	12.3	11.3	10.4	9.8	9.2	8.7	8.3	8.0	7.6	7.4	7.1	6.9	6.7	6.5	6.3	6.2	5.5	5.0	4.4	3.9	6
9	16.6	14.9	13.6	12.6	11.7	11.1	10.5	10.0	9.6	9.2	8.9	8.6	8.3	8.1	7.8	7.6	7.4	6.6	6.1	5.3	4.7	9
12	19.0	17.0	15.5	14.4	13.4	12.7	12.0	11.5	11.0	10.5	10.2	9.8	9.5	9.2	9.0	8.7	8.5	7.6	6.9	6.0	5.4	12
15	21.1	18.9	17.3	16.0	14.9	14.1	13.4	12.7	12.2	11.7	11.3	10.9	10.6	10.3	10.0	9.7	9.5	8.5	7.7	6.7	6.0	15
18	23.1	20.6	18.8	17.4	16.3	15.4	14.6	13.9	13.3	12.8	12.3	11.9	11.5	11.2	10.9	10.6	10.3	9.2	8.4	7.3	6.5	18
21	24.9	22.2	20.3	18.8	17.6	16.6	15.7	15.0	14.4	13.8	13.3	12.8	12.4	12.1	11.7	11.4	11.1	9.9	9.1	7.9	7.0	21
24	26.5	23.7	21.7	20.1	18.8	17.7	16.8	16.0	15.3	14.7	14.2	13.7	13.3	12.9	12.5	12.2	11.9	10.6	9.7	8.4	7.5	24
27	28.1	25.1	22.9	21.2	19.9	18.7	17.8	16.9	16.2	15.6	15.0	14.5	14.0	13.6	13.2	12.9	12.6	11.2	10.3	8.9	7.9	27
30	29.6	26.5	24.2	22.4	20.9	19.7	18.7	17.8	17.1	16.4	15.8	15.3	14.8	14.3	13.9	13.6	13.2	11.8	10.8	9.4	8.4	30
40	34.1	30.5	27.8	25.7	24.1	22.7	21.5	20.5	19.7	18.9	18.2	17.6	17.0	16.5	16.1	15.6	15.2	13.6	12.4	10.8	9.6	40
50	38.0	34.0	31.0	28.7	26.9	25.3	24.0	22.9	22.0	21.1	20.3	19.6	19.0	18.4	17.9	17.4	17.0	15.2	13.9	12.0	10.8	50
60	41.6	37.2	34.0	31.4	29.4	27.7	26.3	25.1	24.0	23.1	22.2	21.5	20.8	20.2	19.6	19.1	18.6	16.6	15.2	13.2	11.8	60
70	44.9	40.2	36.7	33.9	31.7	29.9	28.4	27.1	25.9	24.9	24.0	23.2	22.4	21.8	21.2	20.6	20.1	18.0	16.4	14.2	12.7	70
80	48.0	42.9	39.2	36.3	33.9	32.0	30.3	28.9	27.7	26.6	25.6	24.8	24.0	23.3	22.6	22.0	21.5	19.2	17.5	15.2	13.6	80
90	50.9	45.5	41.5	38.4	36.0	33.9	32.2	30.7	29.4	28.2	27.2	26.3	25.4	24.7	24.0	23.3	22.7	20.3	18.6	16.1	14.4	90
100	53.6	47.9	43.8	40.5	37.9	35.7	33.9	32.3	30.9	29.7	28.6	27.7	26.8	26.0	25.3	24.6	24.0	21.4	19.6	16.9	15.2	100

0.34	4	5	6	7	8	9	10	11	12	13	14	15	16	17	18	19	20	25	30	40	50	1.70
	8 inches past hole																					
3	9.6	8.6	7.9	7.3	6.8	6.4	6.1	5.8	5.6	5.3	5.1	5.0	4.8	4.7	4.5	4.4	4.3	3.9	3.5	3.0	2.7	3
6	13.0	11.6	10.6	9.8	9.2	8.7	8.2	7.8	7.5	7.2	6.9	6.7	6.5	6.3	6.1	6.0	5.8	5.2	4.7	4.1	3.7	6
9	15.6	14.0	12.8	11.8	11.1	10.4	9.9	9.4	9.0	8.7	8.4	8.1	7.8	7.6	7.4	7.2	7.0	6.3	5.7	4.9	4.4	9
12	17.9	16.0	14.6	13.5	12.7	11.9	11.3	10.8	10.3	9.9	9.6	9.2	8.9	8.7	8.4	8.2	8.0	7.2	6.5	5.7	5.1	12
15	19.9	17.8	16.2	15.0	14.1	13.3	12.6	12.0	11.5	11.0	10.6	10.3	9.9	9.7	9.4	9.1	8.9	8.0	7.3	6.3	5.6	15
18	21.7	19.4	17.7	16.4	15.4	14.5	13.7	13.1	12.5	12.0	11.6	11.2	10.9	10.5	10.2	10.0	9.7	8.7	7.9	6.9	6.1	18
21	23.4	20.9	19.1	17.7	16.5	15.6	14.8	14.1	13.5	13.0	12.5	12.1	11.7	11.4	11.0	10.7	10.5	9.4	8.5	7.4	6.6	21
24	25.0	22.3	20.4	18.9	17.7	16.6	15.8	15.1	14.4	13.8	13.3	12.9	12.5	12.1	11.8	11.5	11.2	10.0	9.1	7.9	7.1	24
27	26.4	23.6	21.6	20.0	18.7	17.6	16.7	15.9	15.3	14.7	14.1	13.7	13.2	12.8	12.5	12.1	11.8	10.6	9.7	8.4	7.5	27
30	27.8	24.9	22.7	21.0	19.7	18.6	17.6	16.8	16.1	15.4	14.9	14.4	13.9	13.5	13.1	12.8	12.4	11.1	10.2	8.8	7.9	30
40	32.1	28.7	26.2	24.2	22.7	21.4	20.3	19.3	18.5	17.8	17.1	16.6	16.0	15.6	15.1	14.7	14.3	12.8	11.7	10.1	9.1	40
50	35.8	32.0	29.2	27.0	25.3	23.9	22.6	21.6	20.7	19.8	19.1	18.5	17.9	17.4	16.9	16.4	16.0	14.3	13.1	11.3	10.1	50
60	39.2	35.0	32.0	29.6	27.7	26.1	24.8	23.6	22.6	21.7	20.9	20.2	19.6	19.0	18.5	18.0	17.5	15.7	14.3	12.4	11.1	60
70	42.3	37.8	34.5	31.9	29.9	28.2	26.7	25.5	24.4	23.4	22.6	21.8	21.1	20.5	19.9	19.4	18.9	16.9	15.4	13.4	12.0	70
80	45.1	40.4	36.9	34.1	31.9	30.1	28.6	27.2	26.1	25.0	24.1	23.3	22.6	21.9	21.3	20.7	20.2	18.1	16.5	14.3	12.8	80
90	47.9	42.8	39.1	36.2	33.8	31.9	30.3	28.9	27.6	26.6	25.6	24.7	23.9	23.2	22.6	22.0	21.4	19.1	17.5	15.1	13.5	90
100	50.4	45.1	41.2	38.1	35.7	33.6	31.9	30.4	29.1	28.0	27.0	26.0	25.2	24.5	23.8	23.1	22.6	20.2	18.4	15.9	14.3	100

0.36	4	5	6	7	8	9	10	11	12	13	14	15	16	17	18	19	20	25	30	40	50	1.70
	8 inches past hole																					
3	9.1	8.1	7.4	6.9	6.4	6.1	5.7	5.5	5.2	5.0	4.9	4.7	4.5	4.4	4.3	4.2	4.1	3.6	3.3	2.9	2.6	3
6	12.3	11.0	10.0	9.3	8.7	8.2	7.8	7.4	7.1	6.8	6.6	6.3	6.1	5.9	5.8	5.6	5.5	4.9	4.5	3.9	3.5	6
9	14.8	13.2	12.1	11.2	10.4	9.8	9.3	8.9	8.5	8.2	7.9	7.6	7.4	7.2	7.0	6.8	6.6	5.9	5.4	4.7	4.2	9
12	16.9	15.1	13.8	12.8	11.9	11.3	10.7	10.2	9.8	9.4	9.0	8.7	8.4	8.2	8.0	7.8	7.6	6.8	6.2	5.3	4.8	12
15	18.8	16.8	15.3	14.2	13.3	12.5	11.9	11.3	10.8	10.4	10.0	9.7	9.4	9.1	8.9	8.6	8.4	7.5	6.9	5.9	5.3	15
18	20.5	18.3	16.7	15.5	14.5	13.7	13.0	12.4	11.8	11.4	11.0	10.6	10.3	9.9	9.7	9.4	9.2	8.2	7.5	6.5	5.8	18
21	22.1	19.8	18.0	16.7	15.6	14.7	14.0	13.3	12.8	12.3	11.8	11.4	11.0	10.7	10.4	10.1	9.9	8.8	8.1	7.0	6.3	21
24	23.6	21.1	19.3	17.8	16.7	15.7	14.9	14.2	13.6	13.1	12.6	12.2	11.8	11.4	11.1	10.8	10.5	9.4	8.6	7.5	6.7	24
27	25.0	22.3	20.4	18.9	17.7	16.6	15.8	15.1	14.4	13.9	13.3	12.9	12.5	12.1	11.8	11.5	11.2	10.0	9.1	7.9	7.1	27
30	26.3	23.5	21.5	19.9	18.6	17.5	16.6	15.9	15.2	14.6	14.1	13.6	13.1	12.8	12.4	12.1	11.8	10.5	9.6	8.3	7.4	30
40	30.3	27.1	24.7	22.9	21.4	20.2	19.1	18.3	17.5	16.8	16.2	15.6	15.1	14.7	14.3	13.9	13.5	12.1	11.1	9.6	8.6	40
50	33.8	30.2	27.6	25.5	23.9	22.5	21.4	20.4	19.5	18.7	18.1	17.5	16.9	16.4	15.9	15.5	15.1	13.5	12.3	10.7	9.6	50
60	37.0	33.1	30.2	28.0	26.1	24.7	23.4	22.3	21.3	20.5	19.8	19.1	18.5	17.9	17.4	17.0	16.5	14.8	13.5	11.7	10.5	60
70	39.9	35.7	32.6	30.2	28.2	26.6	25.2	24.1	23.0	22.1	21.3	20.6	20.0	19.4	18.8	18.3	17.8	16.0	14.6	12.6	11.3	70
80	42.6	38.1	34.8	32.2	30.2	28.4	27.0	25.7	24.6	23.7	22.8	22.0	21.3	20.7	20.1	19.6	19.1	17.1	15.6	13.5	12.1	80
90	45.2	40.4	36.9	34.2	32.0	30.1	28.6	27.3	26.1	25.1	24.2	23.3	22.6	21.9	21.3	20.7	20.2	18.1	16.5	14.3	12.8	90
100	47.6	42.6	38.9	36.0	33.7	31.8	30.1	28.7	27.5	26.4	25.5	24.6	23.8	23.1	22.5	21.9	21.3	19.1	17.4	15.1	13.5	100

1.70 Impact Ratio

Pmph/BS″

0.38	4	5	6	7	8	9	10	11	12	13	14	15	16	17	18	19	20	25	30	40	50	1.70
	8 inches past hole																					
3	8.6	7.7	7.0	6.5	6.1	5.7	5.4	5.2	5.0	4.8	4.6	4.4	4.3	4.2	4.1	4.0	3.9	3.4	3.1	2.7	2.4	3
6	11.6	10.4	9.5	8.8	8.2	7.7	7.3	7.0	6.7	6.4	6.2	6.0	5.8	5.6	5.5	5.3	5.2	4.6	4.2	3.7	3.3	6
9	14.0	12.5	11.4	10.6	9.9	9.3	8.8	8.4	8.1	7.8	7.5	7.2	7.0	6.8	6.6	6.4	6.3	5.6	5.1	4.4	4.0	9
12	16.0	14.3	13.1	12.1	11.3	10.7	10.1	9.7	9.2	8.9	8.6	8.3	8.0	7.8	7.5	7.3	7.2	6.4	5.8	5.1	4.5	12
15	17.8	15.9	14.5	13.5	12.6	11.9	11.3	10.7	10.3	9.9	9.5	9.2	8.9	8.6	8.4	8.2	8.0	7.1	6.5	5.6	5.0	15
18	19.4	17.4	15.9	14.7	13.7	13.0	12.3	11.7	11.2	10.8	10.4	10.0	9.7	9.4	9.2	8.9	8.7	7.8	7.1	6.1	5.5	18
21	20.9	18.7	17.1	15.8	14.8	14.0	13.2	12.6	12.1	11.6	11.2	10.8	10.5	10.2	9.9	9.6	9.4	8.4	7.6	6.6	5.9	21
24	22.3	20.0	18.2	16.9	15.8	14.9	14.1	13.5	12.9	12.4	11.9	11.5	11.2	10.8	10.5	10.2	10.0	8.9	8.2	7.1	6.3	24
27	23.7	21.2	19.3	17.9	16.7	15.8	15.0	14.3	13.7	13.1	12.6	12.2	11.8	11.5	11.2	10.9	10.6	9.5	8.6	7.5	6.7	27
30	24.9	22.3	20.3	18.8	17.6	16.6	15.8	15.0	14.4	13.8	13.3	12.9	12.5	12.1	11.7	11.4	11.1	10.0	9.1	7.9	7.0	30
40	28.7	25.7	23.4	21.7	20.3	19.1	18.1	17.3	16.6	15.9	15.3	14.8	14.3	13.9	13.5	13.2	12.8	11.5	10.5	9.1	8.1	40
50	32.0	28.6	26.1	24.2	22.6	21.3	20.2	19.3	18.5	17.8	17.1	16.5	16.0	15.5	15.1	14.7	14.3	12.8	11.7	10.1	9.1	50
60	35.0	31.3	28.6	26.5	24.8	23.4	22.2	21.1	20.2	19.4	18.7	18.1	17.5	17.0	16.5	16.1	15.7	14.0	12.8	11.1	9.9	60
70	37.8	33.8	30.9	28.6	26.7	25.2	23.9	22.8	21.8	21.0	20.2	19.5	18.9	18.3	17.8	17.3	16.9	15.1	13.8	12.0	10.7	70
80	40.4	36.1	33.0	30.5	28.6	26.9	25.5	24.4	23.3	22.4	21.6	20.9	20.2	19.6	19.0	18.5	18.1	16.2	14.8	12.8	11.4	80
90	42.8	38.3	35.0	32.4	30.3	28.6	27.1	25.8	24.7	23.8	22.9	22.1	21.4	20.8	20.2	19.7	19.2	17.1	15.6	13.5	12.1	90
100	45.1	40.4	36.8	34.1	31.9	30.1	28.5	27.2	26.1	25.0	24.1	23.3	22.6	21.9	21.3	20.7	20.2	18.1	16.5	14.3	12.8	100

0.40	4	5	6	7	8	9	10	11	12	13	14	15	16	17	18	19	20	25	30	40	50	1.70
	8 inches past hole																					
3	8.2	7.3	6.7	6.2	5.8	5.5	5.2	4.9	4.7	4.5	4.4	4.2	4.1	4.0	3.9	3.8	3.7	3.3	3.0	2.6	2.3	3
6	11.0	9.9	9.0	8.3	7.8	7.4	7.0	6.7	6.4	6.1	5.9	5.7	5.5	5.4	5.2	5.1	4.9	4.4	4.0	3.5	3.1	6
9	13.3	11.9	10.8	10.0	9.4	8.9	8.4	8.0	7.7	7.4	7.1	6.9	6.6	6.4	6.3	6.1	5.9	5.3	4.9	4.2	3.8	9
12	15.2	13.6	12.4	11.5	10.8	10.1	9.6	9.2	8.8	8.4	8.1	7.9	7.6	7.4	7.2	7.0	6.8	6.1	5.6	4.8	4.3	12
15	16.9	15.1	13.8	12.8	12.0	11.3	10.7	10.2	9.8	9.4	9.0	8.7	8.5	8.2	8.0	7.8	7.6	6.8	6.2	5.3	4.8	15
18	18.5	16.5	15.1	14.0	13.1	12.3	11.7	11.1	10.7	10.2	9.9	9.5	9.2	9.0	8.7	8.5	8.3	7.4	6.7	5.8	5.2	18
21	19.9	17.8	16.2	15.0	14.1	13.3	12.6	12.0	11.5	11.0	10.6	10.3	9.9	9.6	9.4	9.1	8.9	8.0	7.3	6.3	5.6	21
24	21.2	19.0	17.3	16.0	15.0	14.1	13.4	12.8	12.3	11.8	11.3	11.0	10.6	10.3	10.0	9.7	9.5	8.5	7.7	6.7	6.0	24
27	22.5	20.1	18.4	17.0	15.9	15.0	14.2	13.6	13.0	12.5	12.0	11.6	11.2	10.9	10.6	10.3	10.1	9.0	8.2	7.1	6.4	27
30	23.7	21.2	19.3	17.9	16.7	15.8	15.0	14.3	13.7	13.1	12.6	12.2	11.8	11.5	11.2	10.9	10.6	9.5	8.6	7.5	6.7	30
40	27.2	24.4	22.2	20.6	19.3	18.2	17.2	16.4	15.7	15.1	14.6	14.1	13.6	13.2	12.8	12.5	12.2	10.9	9.9	8.6	7.7	40
50	30.4	27.2	24.8	23.0	21.5	20.3	19.2	18.3	17.6	16.9	16.3	15.7	15.2	14.8	14.3	14.0	13.6	12.2	11.1	9.6	8.6	50
60	33.3	29.8	27.2	25.2	23.5	22.2	21.0	20.1	19.2	18.5	17.8	17.2	16.6	16.1	15.7	15.3	14.9	13.3	12.2	10.5	9.4	60
70	35.9	32.1	29.3	27.2	25.4	23.9	22.7	21.7	20.7	19.9	19.2	18.5	18.0	17.4	16.9	16.5	16.1	14.4	13.1	11.4	10.2	70
80	38.4	34.3	31.3	29.0	27.1	25.6	24.3	23.1	22.2	21.3	20.5	19.8	19.2	18.6	18.1	17.6	17.2	15.4	14.0	12.1	10.9	80
90	40.7	36.4	33.2	30.8	28.8	27.1	25.7	24.5	23.5	22.6	21.7	21.0	20.3	19.7	19.2	18.7	18.2	16.3	14.9	12.9	11.5	90
100	42.9	38.3	35.0	32.4	30.3	28.6	27.1	25.9	24.8	23.8	22.9	22.1	21.4	20.8	20.2	19.7	19.2	17.1	15.7	13.6	12.1	100

0.42	4	5	6	7	8	9	10	11	12	13	14	15	16	17	18	19	20	25	30	40	50	1.70
	8 inches past hole																					
3	7.8	7.0	6.4	5.9	5.5	5.2	4.9	4.7	4.5	4.3	4.2	4.0	3.9	3.8	3.7	3.6	3.5	3.1	2.8	2.5	2.2	3
6	10.5	9.4	8.6	7.9	7.4	7.0	6.6	6.3	6.1	5.8	5.6	5.4	5.3	5.1	5.0	4.8	4.7	4.2	3.8	3.3	3.0	6
9	12.7	11.3	10.3	9.6	8.9	8.4	8.0	7.6	7.3	7.0	6.8	6.5	6.3	6.1	6.0	5.8	5.7	5.1	4.6	4.0	3.6	9
12	14.5	13.0	11.8	10.9	10.2	9.7	9.2	8.7	8.4	8.0	7.7	7.5	7.2	7.0	6.8	6.6	6.5	5.8	5.3	4.6	4.1	12
15	16.1	14.4	13.2	12.2	11.4	10.7	10.2	9.7	9.3	8.9	8.6	8.3	8.1	7.8	7.6	7.4	7.2	6.4	5.9	5.1	4.6	15
18	17.6	15.7	14.4	13.3	12.4	11.7	11.1	10.6	10.2	9.8	9.4	9.1	8.8	8.5	8.3	8.1	7.9	7.0	6.4	5.6	5.0	18
21	18.9	16.9	15.5	14.3	13.4	12.6	12.0	11.4	10.9	10.5	10.1	9.8	9.5	9.2	8.9	8.7	8.5	7.6	6.9	6.0	5.4	21
24	20.2	18.1	16.5	15.3	14.3	13.5	12.8	12.2	11.7	11.2	10.8	10.4	10.1	9.8	9.5	9.3	9.0	8.1	7.4	6.4	5.7	24
27	21.4	19.1	17.5	16.2	15.1	14.3	13.5	12.9	12.4	11.9	11.4	11.1	10.7	10.4	10.1	9.8	9.6	8.6	7.8	6.8	6.1	27
30	22.5	20.2	18.4	17.0	15.9	15.0	14.3	13.6	13.0	12.5	12.0	11.6	11.3	10.9	10.6	10.3	10.1	9.0	8.2	7.1	6.4	30
40	26.0	23.2	21.2	19.6	18.4	17.3	16.4	15.6	15.0	14.4	13.9	13.4	13.0	12.6	12.2	11.9	11.6	10.4	9.5	8.2	7.3	40
50	29.0	25.9	23.7	21.9	20.5	19.3	18.3	17.5	16.7	16.1	15.5	15.0	14.5	14.1	13.7	13.3	13.0	11.6	10.6	9.2	8.2	50
60	31.7	28.4	25.9	24.0	22.4	21.1	20.0	19.1	18.3	17.6	16.9	16.4	15.8	15.4	14.9	14.5	14.2	12.7	11.6	10.0	9.0	60
70	34.2	30.6	27.9	25.9	24.2	22.8	21.6	20.6	19.8	19.0	18.3	17.7	17.1	16.6	16.1	15.7	15.3	13.7	12.5	10.8	9.7	70
80	36.5	32.7	29.8	27.6	25.8	24.4	23.1	22.0	21.1	20.3	19.5	18.9	18.3	17.7	17.2	16.8	16.3	14.6	13.3	11.6	10.3	80
90	38.7	34.7	31.6	29.3	27.4	25.8	24.5	23.4	22.4	21.5	20.7	20.0	19.4	18.8	18.3	17.8	17.3	15.5	14.1	12.3	11.0	90
100	40.8	36.5	33.3	30.9	28.9	27.2	25.8	24.6	23.6	22.6	21.8	21.1	20.4	19.8	19.2	18.7	18.3	16.3	14.9	12.9	11.5	100

1.75 Impact Ratio

Pmph/BS"

0.32	4	5	6	7	8	9	10	11	12	13	14	15	16	17	18	19	20	25	30	40	50	1.75
8 inches past hole																						
3	9.9	8.9	8.1	7.5	7.0	6.6	6.3	6.0	5.7	5.5	5.3	5.1	5.0	4.8	4.7	4.6	4.4	4.0	3.6	3.1	2.8	3
6	13.4	12.0	10.9	10.1	9.5	8.9	8.5	8.1	7.7	7.4	7.2	6.9	6.7	6.5	6.3	6.1	6.0	5.4	4.9	4.2	3.8	6
9	16.1	14.4	13.2	12.2	11.4	10.8	10.2	9.7	9.3	8.9	8.6	8.3	8.1	7.8	7.6	7.4	7.2	6.5	5.9	5.1	4.6	9
12	18.5	16.5	15.1	14.0	13.1	12.3	11.7	11.1	10.7	10.2	9.9	9.5	9.2	9.0	8.7	8.5	8.3	7.4	6.7	5.8	5.2	12
15	20.5	18.4	16.8	15.5	14.5	13.7	13.0	12.4	11.9	11.4	11.0	10.6	10.3	10.0	9.7	9.4	9.2	8.2	7.5	6.5	5.8	15
18	22.4	20.1	18.3	16.9	15.9	14.9	14.2	13.5	12.9	12.4	12.0	11.6	11.2	10.9	10.6	10.3	10.0	9.0	8.2	7.1	6.3	18
21	24.2	21.6	19.7	18.3	17.1	16.1	15.3	14.6	13.9	13.4	12.9	12.5	12.1	11.7	11.4	11.1	10.8	9.7	8.8	7.6	6.8	21
24	25.8	23.0	21.0	19.5	18.2	17.2	16.3	15.5	14.9	14.3	13.8	13.3	12.9	12.5	12.1	11.8	11.5	10.3	9.4	8.1	7.3	24
27	27.3	24.4	22.3	20.6	19.3	18.2	17.3	16.5	15.8	15.1	14.6	14.1	13.6	13.2	12.9	12.5	12.2	10.9	10.0	8.6	7.7	27
30	28.7	25.7	23.5	21.7	20.3	19.2	18.2	17.3	16.6	15.9	15.4	14.8	14.4	13.9	13.5	13.2	12.8	11.5	10.5	9.1	8.1	30
40	33.1	29.6	27.0	25.0	23.4	22.1	20.9	20.0	19.1	18.4	17.7	17.1	16.5	16.0	15.6	15.2	14.8	13.2	12.1	10.5	9.4	40
50	36.9	33.0	30.2	27.9	26.1	24.6	23.4	22.3	21.3	20.5	19.7	19.1	18.5	17.9	17.4	16.9	16.5	14.8	13.5	11.7	10.4	50
60	40.4	36.1	33.0	30.5	28.6	26.9	25.6	24.4	23.3	22.4	21.6	20.9	20.2	19.6	19.1	18.5	18.1	16.2	14.8	12.8	11.4	60
70	43.6	39.0	35.6	33.0	30.8	29.1	27.6	26.3	25.2	24.2	23.3	22.5	21.8	21.2	20.6	20.0	19.5	17.4	15.9	13.8	12.3	70
80	46.6	41.7	38.0	35.2	33.0	31.1	29.5	28.1	26.9	25.8	24.9	24.1	23.3	22.6	22.0	21.4	20.8	18.6	17.0	14.7	13.2	80
90	49.4	44.2	40.3	37.3	34.9	32.9	31.2	29.8	28.5	27.4	26.4	25.5	24.7	24.0	23.3	22.7	22.1	19.8	18.0	15.6	14.0	90
100	52.1	46.6	42.5	39.4	36.8	34.7	32.9	31.4	30.1	28.9	27.8	26.9	26.0	25.3	24.5	23.9	23.3	20.8	19.0	16.5	14.7	100

0.34	4	5	6	7	8	9	10	11	12	13	14	15	16	17	18	19	20	25	30	40	50	1.75
8 inches past hole																						
3	9.4	8.4	7.6	7.1	6.6	6.2	5.9	5.6	5.4	5.2	5.0	4.8	4.7	4.5	4.4	4.3	4.2	3.7	3.4	3.0	2.6	3
6	12.6	11.3	10.3	9.5	8.9	8.4	8.0	7.6	7.3	7.0	6.7	6.5	6.3	6.1	5.9	5.8	5.6	5.0	4.6	4.0	3.6	6
9	15.2	13.6	12.4	11.5	10.7	10.1	9.6	9.2	8.8	8.4	8.1	7.8	7.6	7.4	7.2	7.0	6.8	6.1	5.5	4.8	4.3	9
12	17.4	15.5	14.2	13.1	12.3	11.6	11.0	10.5	10.0	9.6	9.3	9.0	8.7	8.4	8.2	8.0	7.8	7.0	6.3	5.5	4.9	12
15	19.3	17.3	15.8	14.6	13.7	12.9	12.2	11.7	11.2	10.7	10.3	10.0	9.7	9.4	9.1	8.9	8.6	7.7	7.1	6.1	5.5	15
18	21.1	18.9	17.2	15.9	14.9	14.1	13.3	12.7	12.2	11.7	11.3	10.9	10.5	10.2	9.9	9.7	9.4	8.4	7.7	6.7	6.0	18
21	22.7	20.3	18.6	17.2	16.1	15.2	14.4	13.7	13.1	12.6	12.2	11.7	11.4	11.0	10.7	10.4	10.2	9.1	8.3	7.2	6.4	21
24	24.3	21.7	19.8	18.3	17.1	16.2	15.3	14.6	14.0	13.5	13.0	12.5	12.1	11.8	11.4	11.1	10.8	9.7	8.9	7.7	6.9	24
27	25.7	23.0	21.0	19.4	18.2	17.1	16.2	15.5	14.8	14.2	13.7	13.3	12.8	12.5	12.1	11.8	11.5	10.3	9.4	8.1	7.3	27
30	27.0	24.2	22.1	20.4	19.1	18.0	17.1	16.3	15.6	15.0	14.5	14.0	13.5	13.1	12.7	12.4	12.1	10.8	9.9	8.6	7.6	30
40	31.1	27.9	25.4	23.5	22.0	20.8	19.7	18.8	18.0	17.3	16.6	16.1	15.6	15.1	14.7	14.3	13.9	12.5	11.4	9.8	8.8	40
50	34.8	31.1	28.4	26.3	24.6	23.2	22.0	21.0	20.1	19.3	18.6	17.9	17.4	16.9	16.4	15.9	15.5	13.9	12.7	11.0	9.8	50
60	38.0	34.0	31.1	28.8	26.9	25.4	24.1	22.9	22.0	21.1	20.3	19.6	19.0	18.5	17.9	17.5	17.0	15.2	13.9	12.0	10.8	60
70	41.1	36.7	33.5	31.0	29.0	27.4	26.0	24.8	23.7	22.8	21.9	21.2	20.5	19.9	19.4	18.8	18.4	16.4	15.0	13.0	11.6	70
80	43.9	39.2	35.8	33.2	31.0	29.2	27.7	26.4	25.3	24.3	23.4	22.6	21.9	21.3	20.7	20.1	19.6	17.5	16.0	13.9	12.4	80
90	46.5	41.6	38.0	35.1	32.9	31.0	29.4	28.0	26.8	25.8	24.9	24.0	23.2	22.6	21.9	21.3	20.8	18.6	17.0	14.7	13.2	90
100	49.0	43.8	40.0	37.0	34.6	32.7	31.0	29.5	28.3	27.2	26.2	25.3	24.5	23.8	23.1	22.5	21.9	19.6	17.9	15.5	13.9	100

0.36	4	5	6	7	8	9	10	11	12	13	14	15	16	17	18	19	20	25	30	40	50	1.75
8 inches past hole																						
3	8.8	7.9	7.2	6.7	6.2	5.9	5.6	5.3	5.1	4.9	4.7	4.6	4.4	4.3	4.2	4.1	3.9	3.5	3.2	2.8	2.5	3
6	11.9	10.7	9.7	9.0	8.4	7.9	7.5	7.2	6.9	6.6	6.4	6.1	6.0	5.8	5.6	5.5	5.3	4.8	4.3	3.8	3.4	6
9	14.3	12.8	11.7	10.8	10.1	9.6	9.1	8.6	8.3	8.0	7.7	7.4	7.2	7.0	6.8	6.6	6.4	5.7	5.2	4.5	4.1	9
12	16.4	14.7	13.4	12.4	11.6	10.9	10.4	9.9	9.5	9.1	8.8	8.5	8.2	8.0	7.7	7.5	7.3	6.6	6.0	5.2	4.6	12
15	18.3	16.3	14.9	13.8	12.9	12.2	11.5	11.0	10.5	10.1	9.8	9.4	9.1	8.9	8.6	8.4	8.2	7.3	6.7	5.8	5.2	15
18	19.9	17.8	16.3	15.1	14.1	13.3	12.6	12.0	11.5	11.1	10.7	10.3	10.0	9.7	9.4	9.1	8.9	8.0	7.3	6.3	5.6	18
21	21.5	19.2	17.5	16.2	15.2	14.3	13.6	12.9	12.4	11.9	11.5	11.1	10.7	10.4	10.1	9.9	9.6	8.6	7.8	6.8	6.1	21
24	22.9	20.5	18.7	17.3	16.2	15.3	14.5	13.8	13.2	12.7	12.2	11.8	11.5	11.1	10.8	10.5	10.2	9.2	8.4	7.2	6.5	24
27	24.3	21.7	19.8	18.3	17.2	16.2	15.3	14.6	14.0	13.5	13.0	12.5	12.1	11.8	11.4	11.1	10.8	9.7	8.9	7.7	6.9	27
30	25.5	22.8	20.9	19.3	18.1	17.0	16.2	15.4	14.7	14.2	13.7	13.2	12.8	12.4	12.0	11.7	11.4	10.2	9.3	8.1	7.2	30
40	29.4	26.3	24.0	22.2	20.8	19.6	18.6	17.7	17.0	16.3	15.7	15.2	14.7	14.3	13.9	13.5	13.2	11.8	10.7	9.3	8.3	40
50	32.8	29.4	26.8	24.8	23.2	21.9	20.8	19.8	19.0	18.2	17.5	17.0	16.4	15.9	15.5	15.1	14.7	13.1	12.0	10.4	9.3	50
60	35.9	32.1	29.3	27.2	25.4	23.9	22.7	21.7	20.7	19.9	19.2	18.6	18.0	17.4	16.9	16.5	16.1	14.4	13.1	11.4	10.2	60
70	38.8	34.7	31.7	29.3	27.4	25.8	24.5	23.4	22.4	21.5	20.7	20.0	19.4	18.8	18.3	17.8	17.3	15.5	14.2	12.3	11.0	70
80	41.4	37.0	33.8	31.3	29.3	27.6	26.2	25.0	23.9	23.0	22.1	21.4	20.7	20.1	19.5	19.0	18.5	16.6	15.1	13.1	11.7	80
90	43.9	39.3	35.9	33.2	31.1	29.3	27.8	26.5	25.4	24.4	23.5	22.7	22.0	21.3	20.7	20.1	19.6	17.6	16.0	13.9	12.4	90
100	46.3	41.4	37.8	35.0	32.7	30.8	29.3	27.9	26.7	25.7	24.7	23.9	23.1	22.4	21.8	21.2	20.7	18.5	16.9	14.6	13.1	100

1.75 Impact Ratio

Pmph/BS"	0.38	4	5	6	7	8	9	10	11	12	13	14	15	16	17	18	19	20	25	30	40	50	1.75	Impact Ratio
		8 inches past hole																						
	3	8.4	7.5	6.8	6.3	5.9	5.6	5.3	5.0	4.8	4.6	4.5	4.3	4.2	4.1	3.9	3.8	3.7	3.3	3.1	2.6	2.4	3	
	6	11.3	10.1	9.2	8.5	8.0	7.5	7.1	6.8	6.5	6.3	6.0	5.8	5.6	5.5	5.3	5.2	5.0	4.5	4.1	3.6	3.2	6	
	9	13.6	12.2	11.1	10.3	9.6	9.1	8.6	8.2	7.8	7.5	7.3	7.0	6.8	6.6	6.4	6.2	6.1	5.4	5.0	4.3	3.8	9	
	12	15.6	13.9	12.7	11.8	11.0	10.4	9.8	9.4	9.0	8.6	8.3	8.0	7.8	7.5	7.3	7.1	7.0	6.2	5.7	4.9	4.4	12	
	15	17.3	15.5	14.1	13.1	12.2	11.5	10.9	10.4	10.0	9.6	9.2	8.9	8.6	8.4	8.2	7.9	7.7	6.9	6.3	5.5	4.9	15	
	18	18.9	16.9	15.4	14.3	13.3	12.6	11.9	11.4	10.9	10.5	10.1	9.7	9.4	9.2	8.9	8.7	8.4	7.6	6.9	6.0	5.3	18	
	21	20.3	18.2	16.6	15.4	14.4	13.6	12.9	12.3	11.7	11.3	10.9	10.5	10.2	9.9	9.6	9.3	9.1	8.1	7.4	6.4	5.8	21	
	24	21.7	19.4	17.7	16.4	15.3	14.5	13.7	13.1	12.5	12.0	11.6	11.2	10.9	10.5	10.2	10.0	9.7	8.7	7.9	6.9	6.1	24	
	27	23.0	20.6	18.8	17.4	16.3	15.3	14.5	13.9	13.3	12.7	12.3	11.9	11.5	11.1	10.8	10.5	10.3	9.2	8.4	7.3	6.5	27	
	30	24.2	21.6	19.8	18.3	17.1	16.1	15.3	14.6	14.0	13.4	12.9	12.5	12.1	11.7	11.4	11.1	10.8	9.7	8.8	7.7	6.8	30	
	40	27.9	24.9	22.8	21.1	19.7	18.6	17.6	16.8	16.1	15.5	14.9	14.4	13.9	13.5	13.1	12.8	12.5	11.1	10.2	8.8	7.9	40	
	50	31.1	27.8	25.4	23.5	22.0	20.7	19.7	18.8	18.0	17.3	16.6	16.1	15.6	15.1	14.7	14.3	13.9	12.4	11.4	9.8	8.8	50	
	60	34.0	30.4	27.8	25.7	24.1	22.7	21.5	20.5	19.6	18.9	18.2	17.6	17.0	16.5	16.0	15.6	15.2	13.6	12.4	10.8	9.6	60	
	70	36.7	32.9	30.0	27.8	26.0	24.5	23.2	22.1	21.2	20.4	19.6	19.0	18.4	17.8	17.3	16.9	16.4	14.7	13.4	11.6	10.4	70	
	80	39.2	35.1	32.0	29.7	27.7	26.2	24.8	23.7	22.7	21.8	21.0	20.3	19.6	19.0	18.5	18.0	17.5	15.7	14.3	12.4	11.1	80	
	90	41.6	37.2	34.0	31.4	29.4	27.7	26.3	25.1	24.0	23.1	22.2	21.5	20.8	20.2	19.6	19.1	18.6	16.6	15.2	13.2	11.8	90	
	100	43.8	39.2	35.8	33.1	31.0	29.2	27.7	26.4	25.3	24.3	23.4	22.6	21.9	21.3	20.7	20.1	19.6	17.5	16.0	13.9	12.4	100	

	0.40	4	5	6	7	8	9	10	11	12	13	14	15	16	17	18	19	20	25	30	40	50	1.75
		8 inches past hole																					
	3	7.9	7.1	6.5	6.0	5.6	5.3	5.0	4.8	4.6	4.4	4.2	4.1	4.0	3.9	3.7	3.6	3.6	3.2	2.9	2.5	2.2	3
	6	10.7	9.6	8.8	8.1	7.6	7.1	6.8	6.5	6.2	5.9	5.7	5.5	5.4	5.2	5.1	4.9	4.8	4.3	3.9	3.4	3.0	6
	9	12.9	11.5	10.5	9.8	9.1	8.6	8.2	7.8	7.5	7.2	6.9	6.7	6.5	6.3	6.1	5.9	5.8	5.2	4.7	4.1	3.7	9
	12	14.8	13.2	12.1	11.2	10.4	9.8	9.3	8.9	8.5	8.2	7.9	7.6	7.4	7.2	7.0	6.8	6.6	5.9	5.4	4.7	4.2	12
	15	16.4	14.7	13.4	12.4	11.6	11.0	10.4	9.9	9.5	9.1	8.8	8.5	8.2	8.0	7.7	7.5	7.3	6.6	6.0	5.2	4.6	15
	18	17.9	16.0	14.6	13.6	12.7	12.0	11.3	10.8	10.4	9.9	9.6	9.3	9.0	8.7	8.5	8.2	8.0	7.2	6.5	5.7	5.1	18
	21	19.3	17.3	15.8	14.6	13.7	12.9	12.2	11.7	11.2	10.7	10.3	10.0	9.7	9.4	9.1	8.9	8.6	7.7	7.1	6.1	5.5	21
	24	20.6	18.4	16.8	15.6	14.6	13.7	13.0	12.4	11.9	11.4	11.0	10.6	10.3	10.0	9.7	9.5	9.2	8.2	7.5	6.5	5.8	24
	27	21.8	19.5	17.8	16.5	15.4	14.6	13.8	13.2	12.6	12.1	11.7	11.3	10.9	10.6	10.3	10.0	9.8	8.7	8.0	6.9	6.2	27
	30	23.0	20.6	18.8	17.4	16.3	15.3	14.5	13.9	13.3	12.8	12.3	11.9	11.5	11.1	10.8	10.5	10.3	9.2	8.4	7.3	6.5	30
	40	26.5	23.7	21.6	20.0	18.7	17.6	16.7	16.0	15.3	14.7	14.1	13.7	13.2	12.8	12.5	12.1	11.8	10.6	9.7	8.4	7.5	40
	50	29.5	26.4	24.1	22.3	20.9	19.7	18.7	17.8	17.1	16.4	15.8	15.3	14.8	14.3	13.9	13.6	13.2	11.8	10.8	9.3	8.4	50
	60	32.3	28.9	26.4	24.4	22.9	21.6	20.4	19.5	18.7	17.9	17.3	16.7	16.2	15.7	15.2	14.8	14.5	12.9	11.8	10.2	9.1	60
	70	34.9	31.2	28.5	26.4	24.7	23.3	22.1	21.0	20.1	19.4	18.7	18.0	17.4	16.9	16.4	16.0	15.6	14.0	12.7	11.0	9.9	70
	80	37.3	33.3	30.4	28.2	26.4	24.9	23.6	22.5	21.5	20.7	19.9	19.3	18.6	18.1	17.6	17.1	16.7	14.9	13.6	11.8	10.5	80
	90	39.5	35.4	32.3	29.9	27.9	26.3	25.0	23.8	22.8	21.9	21.1	20.4	19.8	19.2	18.6	18.1	17.7	15.8	14.4	12.5	11.2	90
	100	41.6	37.2	34.0	31.5	29.4	27.8	26.3	25.1	24.0	23.1	22.3	21.5	20.8	20.2	19.6	19.1	18.6	16.7	15.2	13.2	11.8	100

	0.42	4	5	6	7	8	9	10	11	12	13	14	15	16	17	18	19	20	25	30	40	50	1.75
		8 inches past hole																					
	3	7.6	6.8	6.2	5.7	5.4	5.0	4.8	4.6	4.4	4.2	4.0	3.9	3.8	3.7	3.6	3.5	3.4	3.0	2.8	2.4	2.1	3
	6	10.2	9.1	8.3	7.7	7.2	6.8	6.5	6.2	5.9	5.7	5.5	5.3	5.1	5.0	4.8	4.7	4.6	4.1	3.7	3.2	2.9	6
	9	12.3	11.0	10.0	9.3	8.7	8.2	7.8	7.4	7.1	6.8	6.6	6.3	6.1	6.0	5.8	5.6	5.5	4.9	4.5	3.9	3.5	9
	12	14.1	12.6	11.5	10.6	9.9	9.4	8.9	8.5	8.1	7.8	7.5	7.3	7.0	6.8	6.6	6.5	6.3	5.6	5.1	4.4	4.0	12
	15	15.6	14.0	12.8	11.8	11.1	10.4	9.9	9.4	9.0	8.7	8.4	8.1	7.8	7.6	7.4	7.2	7.0	6.3	5.7	4.9	4.4	15
	18	17.1	15.3	13.9	12.9	12.1	11.4	10.8	10.3	9.9	9.5	9.1	8.8	8.5	8.3	8.1	7.8	7.6	6.8	6.2	5.4	4.8	18
	21	18.4	16.5	15.0	13.9	13.0	12.3	11.6	11.1	10.6	10.2	9.8	9.5	9.2	8.9	8.7	8.4	8.2	7.4	6.7	5.8	5.2	21
	24	19.6	17.6	16.0	14.8	13.9	13.1	12.4	11.8	11.3	10.9	10.5	10.1	9.8	9.5	9.3	9.0	8.8	7.9	7.2	6.2	5.6	24
	27	20.8	18.6	17.0	15.7	14.7	13.9	13.2	12.5	12.0	11.5	11.1	10.7	10.4	10.1	9.8	9.5	9.3	8.3	7.6	6.6	5.9	27
	30	21.9	19.6	17.9	16.5	15.5	14.6	13.8	13.2	12.6	12.1	11.7	11.3	10.9	10.6	10.3	10.0	9.8	8.8	8.0	6.9	6.2	30
	40	25.2	22.5	20.6	19.1	17.8	16.8	15.9	15.2	14.6	14.0	13.5	13.0	12.6	12.2	11.9	11.6	11.3	10.1	9.2	8.0	7.1	40
	50	28.1	25.2	23.0	21.3	19.9	18.8	17.8	17.0	16.2	15.6	15.0	14.5	14.1	13.6	13.3	12.9	12.6	11.3	10.3	8.9	8.0	50
	60	30.8	27.5	25.1	23.3	21.8	20.5	19.5	18.6	17.8	17.1	16.5	15.9	15.4	14.9	14.5	14.1	13.8	12.3	11.2	9.7	8.7	60
	70	33.2	29.7	27.1	25.1	23.5	22.2	21.0	20.0	19.2	18.4	17.8	17.2	16.6	16.1	15.7	15.2	14.9	13.3	12.1	10.5	9.4	70
	80	35.5	31.8	29.0	26.8	25.1	23.7	22.5	21.4	20.5	19.7	19.0	18.3	17.8	17.2	16.7	16.3	15.9	14.2	13.0	11.2	10.0	80
	90	37.6	33.7	30.7	28.5	26.6	25.1	23.8	22.7	21.7	20.9	20.1	19.4	18.8	18.3	17.7	17.3	16.8	15.1	13.7	11.9	10.6	90
	100	39.7	35.5	32.4	30.0	28.0	26.4	25.1	23.9	22.9	22.0	21.2	20.5	19.8	19.2	18.7	18.2	17.7	15.9	14.5	12.5	11.2	100

1.80 Impact Ratio

Pmph/BS"

0.32	4	5	6	7	8	9	10	11	12	13	14	15	16	17	18	19	20	25	30	40	50	1.80 Impact Ratio
8 inches past hole																						
3	9.7	8.6	7.9	7.3	6.8	6.4	6.1	5.8	5.6	5.4	5.2	5.0	4.8	4.7	4.6	4.4	4.3	3.9	3.5	3.1	2.7	3
6	13.0	11.6	10.6	9.8	9.2	8.7	8.2	7.9	7.5	7.2	7.0	6.7	6.5	6.3	6.1	6.0	5.8	5.2	4.8	4.1	3.7	6
9	15.7	14.0	12.8	11.9	11.1	10.5	9.9	9.5	9.1	8.7	8.4	8.1	7.8	7.6	7.4	7.2	7.0	6.3	5.7	5.0	4.4	9
12	18.0	16.1	14.7	13.6	12.7	12.0	11.4	10.8	10.4	10.0	9.6	9.3	9.0	8.7	8.5	8.2	8.0	7.2	6.6	5.7	5.1	12
15	20.0	17.9	16.3	15.1	14.1	13.3	12.6	12.0	11.5	11.1	10.7	10.3	10.0	9.7	9.4	9.2	8.9	8.0	7.3	6.3	5.6	15
18	21.8	19.5	17.8	16.5	15.4	14.5	13.8	13.1	12.6	12.1	11.6	11.3	10.9	10.6	10.3	10.0	9.7	8.7	8.0	6.9	6.2	18
21	23.5	21.0	19.2	17.7	16.6	15.7	14.9	14.2	13.6	13.0	12.6	12.1	11.7	11.4	11.1	10.8	10.5	9.4	8.6	7.4	6.6	21
24	25.1	22.4	20.5	18.9	17.7	16.7	15.8	15.1	14.5	13.9	13.4	12.9	12.5	12.2	11.8	11.5	11.2	10.0	9.1	7.9	7.1	24
27	26.5	23.7	21.7	20.1	18.8	17.7	16.8	16.0	15.3	14.7	14.2	13.7	13.3	12.9	12.5	12.2	11.9	10.6	9.7	8.4	7.5	27
30	27.9	25.0	22.8	21.1	19.8	18.6	17.7	16.8	16.1	15.5	14.9	14.4	14.0	13.6	13.2	12.8	12.5	11.2	10.2	8.8	7.9	30
40	32.2	28.8	26.3	24.3	22.7	21.4	20.3	19.4	18.6	17.8	17.2	16.6	16.1	15.6	15.2	14.8	14.4	12.9	11.7	10.2	9.1	40
50	35.9	32.1	29.3	27.1	25.4	23.9	22.7	21.7	20.7	19.9	19.2	18.5	18.0	17.4	16.9	16.5	16.1	14.4	13.1	11.4	10.2	50
60	39.3	35.1	32.1	29.7	27.8	26.2	24.8	23.7	22.7	21.8	21.0	20.3	19.6	19.1	18.5	18.0	17.6	15.7	14.3	12.4	11.1	60
70	42.4	37.9	34.6	32.1	30.0	28.3	26.8	25.6	24.5	23.5	22.7	21.9	21.2	20.6	20.0	19.5	19.0	17.0	15.5	13.4	12.0	70
80	45.3	40.5	37.0	34.2	32.0	30.2	28.7	27.3	26.2	25.1	24.2	23.4	22.7	22.0	21.4	20.8	20.3	18.1	16.5	14.3	12.8	80
90	48.0	43.0	39.2	36.3	34.0	32.0	30.4	29.0	27.7	26.6	25.7	24.8	24.0	23.3	22.6	22.0	21.5	19.2	17.5	15.2	13.6	90
100	50.6	45.3	41.3	38.3	35.8	33.7	32.0	30.5	29.2	28.1	27.1	26.1	25.3	24.6	23.9	23.2	22.6	20.2	18.5	16.0	14.3	100

0.34	4	5	6	7	8	9	10	11	12	13	14	15	16	17	18	19	20	25	30	40	50	1.80
8 inches past hole																						
3	9.1	8.1	7.4	6.9	6.4	6.1	5.7	5.5	5.2	5.0	4.9	4.7	4.5	4.4	4.3	4.2	4.1	3.6	3.3	2.9	2.6	3
6	12.3	11.0	10.0	9.3	8.7	8.2	7.8	7.4	7.1	6.8	6.6	6.3	6.1	5.9	5.8	5.6	5.5	4.9	4.5	3.9	3.5	6
9	14.8	13.2	12.1	11.2	10.4	9.8	9.3	8.9	8.5	8.2	7.9	7.6	7.4	7.2	7.0	6.8	6.6	5.9	5.4	4.7	4.2	9
12	16.9	15.1	13.8	12.8	11.9	11.3	10.7	10.2	9.8	9.4	9.0	8.7	8.4	8.2	8.0	7.8	7.6	6.8	6.2	5.3	4.8	12
15	18.8	16.8	15.3	14.2	13.3	12.5	11.9	11.3	10.8	10.4	10.0	9.7	9.4	9.1	8.9	8.6	8.4	7.5	6.9	5.9	5.3	15
18	20.5	18.3	16.7	15.5	14.5	13.7	13.0	12.4	11.8	11.4	11.0	10.6	10.3	9.9	9.7	9.4	9.2	8.2	7.5	6.5	5.8	18
21	22.1	19.8	18.0	16.7	15.6	14.7	14.0	13.3	12.8	12.3	11.8	11.4	11.0	10.7	10.4	10.1	9.9	8.8	8.1	7.0	6.3	21
24	23.6	21.1	19.3	17.8	16.7	15.7	14.9	14.2	13.6	13.1	12.6	12.2	11.8	11.4	11.1	10.8	10.5	9.4	8.6	7.5	6.7	24
27	25.0	22.3	20.4	18.9	17.7	16.6	15.8	15.1	14.4	13.9	13.3	12.9	12.5	12.1	11.8	11.5	11.2	10.0	9.1	7.9	7.1	27
30	26.3	23.5	21.5	19.9	18.6	17.5	16.6	15.9	15.2	14.6	14.1	13.6	13.1	12.8	12.4	12.1	11.8	10.5	9.6	8.3	7.4	30
40	30.3	27.1	24.7	22.9	21.4	20.2	19.1	18.3	17.5	16.8	16.2	15.6	15.1	14.7	14.3	13.9	13.5	12.1	11.1	9.6	8.6	40
50	33.8	30.2	27.6	25.5	23.9	22.5	21.4	20.4	19.5	18.7	18.1	17.5	16.9	16.4	15.9	15.5	15.1	13.5	12.3	10.7	9.6	50
60	37.0	33.1	30.2	28.0	26.1	24.7	23.4	22.3	21.3	20.5	19.8	19.1	18.5	17.9	17.4	17.0	16.5	14.8	13.5	11.7	10.5	60
70	39.9	35.7	32.6	30.2	28.2	26.6	25.2	24.1	23.0	22.1	21.3	20.6	20.0	19.4	18.8	18.3	17.8	16.0	14.6	12.6	11.3	70
80	42.6	38.1	34.8	32.2	30.2	28.4	27.0	25.7	24.6	23.7	22.8	22.0	21.3	20.7	20.1	19.6	19.1	17.1	15.6	13.5	12.1	80
90	45.2	40.4	36.9	34.2	32.0	30.1	28.6	27.3	26.1	25.1	24.2	23.3	22.6	21.9	21.3	20.7	20.2	18.1	16.5	14.3	12.8	90
100	47.6	42.6	38.9	36.0	33.7	31.8	30.1	28.7	27.5	26.4	25.5	24.6	23.8	23.1	22.5	21.9	21.3	19.1	17.4	15.1	13.5	100

0.36	4	5	6	7	8	9	10	11	12	13	14	15	16	17	18	19	20	25	30	40	50	1.80
8 inches past hole																						
3	8.6	7.7	7.0	6.5	6.1	5.7	5.4	5.2	5.0	4.8	4.6	4.4	4.3	4.2	4.0	3.9	3.8	3.4	3.1	2.7	2.4	3
6	11.6	10.4	9.5	8.8	8.2	7.7	7.3	7.0	6.7	6.4	6.2	6.0	5.8	5.6	5.5	5.3	5.2	4.6	4.2	3.7	3.3	6
9	13.9	12.5	11.4	10.5	9.9	9.3	8.8	8.4	8.0	7.7	7.5	7.2	7.0	6.8	6.6	6.4	6.2	5.6	5.1	4.4	3.9	9
12	16.0	14.3	13.0	12.1	11.3	10.6	10.1	9.6	9.2	8.9	8.5	8.2	8.0	7.7	7.5	7.3	7.1	6.4	5.8	5.0	4.5	12
15	17.7	15.9	14.5	13.4	12.5	11.8	11.2	10.7	10.2	9.8	9.5	9.2	8.9	8.6	8.4	8.1	7.9	7.1	6.5	5.6	5.0	15
18	19.4	17.3	15.8	14.6	13.7	12.9	12.3	11.7	11.2	10.7	10.4	10.0	9.7	9.4	9.1	8.9	8.7	7.7	7.1	6.1	5.5	18
21	20.9	18.7	17.0	15.8	14.8	13.9	13.2	12.6	12.1	11.6	11.2	10.8	10.4	10.1	9.8	9.6	9.3	8.3	7.6	6.6	5.9	21
24	22.3	19.9	18.2	16.8	15.7	14.8	14.1	13.4	12.9	12.4	11.9	11.5	11.1	10.8	10.5	10.2	10.0	8.9	8.1	7.0	6.3	24
27	23.6	21.1	19.3	17.8	16.7	15.7	14.9	14.2	13.6	13.1	12.6	12.2	11.8	11.4	11.1	10.8	10.5	9.4	8.6	7.5	6.7	27
30	24.8	22.2	20.3	18.8	17.6	16.6	15.7	15.0	14.3	13.8	13.3	12.8	12.4	12.0	11.7	11.4	11.1	9.9	9.1	7.9	7.0	30
40	28.6	25.6	23.3	21.6	20.2	19.1	18.1	17.2	16.5	15.9	15.3	14.8	14.3	13.9	13.5	13.1	12.8	11.4	10.4	9.0	8.1	40
50	31.9	28.5	26.1	24.1	22.6	21.3	20.2	19.2	18.4	17.7	17.1	16.5	16.0	15.5	15.0	14.6	14.3	12.8	11.7	10.1	9.0	50
60	34.9	31.2	28.5	26.4	24.7	23.3	22.1	21.1	20.2	19.4	18.7	18.0	17.5	16.9	16.5	16.0	15.6	14.0	12.8	11.0	9.9	60
70	37.7	33.7	30.8	28.5	26.7	25.1	23.8	22.7	21.8	20.9	20.1	19.5	18.8	18.3	17.8	17.3	16.9	15.1	13.8	11.9	10.7	70
80	40.3	36.0	32.9	30.4	28.5	26.8	25.5	24.3	23.3	22.3	21.5	20.8	20.1	19.5	19.0	18.5	18.0	16.1	14.7	12.7	11.4	80
90	42.7	38.2	34.9	32.3	30.2	28.5	27.0	25.7	24.7	23.7	22.8	22.0	21.3	20.7	20.1	19.6	19.1	17.1	15.6	13.5	12.1	90
100	45.0	40.2	36.7	34.0	31.8	30.0	28.5	27.1	26.0	25.0	24.0	23.2	22.5	21.8	21.2	20.6	20.1	18.0	16.4	14.2	12.7	100

1.80 Impact Ratio

Pmph/BS"	0.38	4	5	6	7	8	9	10	11	12	13	14	15	16	17	18	19	20	25	30	40	50	1.80	Impact Ratio
	8 inches past hole																							
	3	8.1	7.3	6.6	6.1	5.8	5.4	5.1	4.9	4.7	4.5	4.3	4.2	4.1	3.9	3.8	3.7	3.6	3.3	3.0	2.6	2.3	3	
	6	11.0	9.8	9.0	8.3	7.8	7.3	6.9	6.6	6.3	6.1	5.9	5.7	5.5	5.3	5.2	5.0	4.9	4.4	4.0	3.5	3.1	6	
	9	13.2	11.8	10.8	10.0	9.3	8.8	8.4	8.0	7.6	7.3	7.1	6.8	6.6	6.4	6.2	6.1	5.9	5.3	4.8	4.2	3.7	9	
	12	15.1	13.5	12.3	11.4	10.7	10.1	9.6	9.1	8.7	8.4	8.1	7.8	7.6	7.3	7.1	6.9	6.8	6.0	5.5	4.8	4.3	12	
	15	16.8	15.0	13.7	12.7	11.9	11.2	10.6	10.1	9.7	9.3	9.0	8.7	8.4	8.2	7.9	7.7	7.5	6.7	6.1	5.3	4.8	15	
	18	18.4	16.4	15.0	13.9	13.0	12.2	11.6	11.1	10.6	10.2	9.8	9.5	9.2	8.9	8.7	8.4	8.2	7.3	6.7	5.8	5.2	18	
	21	19.8	17.7	16.1	14.9	14.0	13.2	12.5	11.9	11.4	11.0	10.6	10.2	9.9	9.6	9.3	9.1	8.8	7.9	7.2	6.3	5.6	21	
	24	21.1	18.9	17.2	15.9	14.9	14.1	13.3	12.7	12.2	11.7	11.3	10.9	10.5	10.2	9.9	9.7	9.4	8.4	7.7	6.7	6.0	24	
	27	22.3	20.0	18.2	16.9	15.8	14.9	14.1	13.5	12.9	12.4	11.9	11.5	11.2	10.8	10.5	10.3	10.0	8.9	8.2	7.1	6.3	27	
	30	23.5	21.0	19.2	17.8	16.6	15.7	14.9	14.2	13.6	13.0	12.6	12.1	11.8	11.4	11.1	10.8	10.5	9.4	8.6	7.4	6.7	30	
	40	27.1	24.2	22.1	20.5	19.2	18.1	17.1	16.3	15.6	15.0	14.5	14.0	13.5	13.1	12.8	12.4	12.1	10.8	9.9	8.6	7.7	40	
	50	30.2	27.0	24.7	22.9	21.4	20.2	19.1	18.2	17.5	16.8	16.2	15.6	15.1	14.7	14.3	13.9	13.5	12.1	11.0	9.6	8.6	50	
	60	33.1	29.6	27.0	25.0	23.4	22.1	20.9	20.0	19.1	18.4	17.7	17.1	16.5	16.0	15.6	15.2	14.8	13.2	12.1	10.5	9.4	60	
	70	35.7	31.9	29.2	27.0	25.3	23.8	22.6	21.5	20.6	19.8	19.1	18.4	17.9	17.3	16.8	16.4	16.0	14.3	13.0	11.3	10.1	70	
	80	38.2	34.1	31.2	28.8	27.0	25.4	24.1	23.0	22.0	21.2	20.4	19.7	19.1	18.5	18.0	17.5	17.1	15.3	13.9	12.1	10.8	80	
	90	40.4	36.2	33.0	30.6	28.6	27.0	25.6	24.4	23.4	22.4	21.6	20.9	20.2	19.6	19.1	18.6	18.1	16.2	14.8	12.8	11.4	90	
	100	42.6	38.1	34.8	32.2	30.1	28.4	27.0	25.7	24.6	23.6	22.8	22.0	21.3	20.7	20.1	19.6	19.1	17.0	15.6	13.5	12.1	100	

	0.40	4	5	6	7	8	9	10	11	12	13	14	15	16	17	18	19	20	25	30	40	50	1.80
	8 inches past hole																						
	3	7.7	6.9	6.3	5.8	5.5	5.2	4.9	4.7	4.5	4.3	4.1	4.0	3.9	3.7	3.6	3.5	3.5	3.1	2.8	2.4	2.2	3
	6	10.4	9.3	8.5	7.9	7.4	6.9	6.6	6.3	6.0	5.8	5.6	5.4	5.2	5.1	4.9	4.8	4.7	4.2	3.8	3.3	2.9	6
	9	12.5	11.2	10.2	9.5	8.9	8.4	7.9	7.6	7.2	7.0	6.7	6.5	6.3	6.1	5.9	5.8	5.6	5.0	4.6	4.0	3.5	9
	12	14.4	12.8	11.7	10.9	10.2	9.6	9.1	8.7	8.3	8.0	7.7	7.4	7.2	7.0	6.8	6.6	6.4	5.7	5.2	4.5	4.1	12
	15	16.0	14.3	13.0	12.1	11.3	10.6	10.1	9.6	9.2	8.9	8.5	8.2	8.0	7.7	7.5	7.3	7.1	6.4	5.8	5.1	4.5	15
	18	17.4	15.6	14.2	13.2	12.3	11.6	11.0	10.5	10.1	9.7	9.3	9.0	8.7	8.5	8.2	8.0	7.8	7.0	6.4	5.5	4.9	18
	21	18.8	16.8	15.3	14.2	13.3	12.5	11.9	11.3	10.8	10.4	10.0	9.7	9.4	9.1	8.9	8.6	8.4	7.5	6.9	5.9	5.3	21
	24	20.0	17.9	16.4	15.2	14.2	13.4	12.7	12.1	11.6	11.1	10.7	10.4	10.0	9.7	9.4	9.2	9.0	8.0	7.3	6.3	5.7	24
	27	21.2	19.0	17.3	16.0	15.0	14.2	13.4	12.8	12.3	11.8	11.3	11.0	10.6	10.3	10.0	9.7	9.5	8.5	7.8	6.7	6.0	27
	30	22.3	20.0	18.2	16.9	15.8	14.9	14.1	13.5	12.9	12.4	11.9	11.5	11.2	10.8	10.5	10.3	10.0	8.9	8.2	7.1	6.3	30
	40	25.7	23.0	21.0	19.5	18.2	17.2	16.3	15.5	14.9	14.3	13.8	13.3	12.9	12.5	12.1	11.8	11.5	10.3	9.4	8.1	7.3	40
	50	28.7	25.7	23.5	21.7	20.3	19.2	18.2	17.3	16.6	15.9	15.4	14.8	14.4	13.9	13.5	13.2	12.8	11.5	10.5	9.1	8.1	50
	60	31.4	28.1	25.7	23.8	22.2	21.0	19.9	19.0	18.1	17.4	16.8	16.2	15.7	15.2	14.8	14.4	14.1	12.6	11.5	9.9	8.9	60
	70	33.9	30.3	27.7	25.6	24.0	22.6	21.5	20.5	19.6	18.8	18.1	17.5	17.0	16.5	16.0	15.6	15.2	13.6	12.4	10.7	9.6	70
	80	36.2	32.4	29.6	27.4	25.6	24.2	22.9	21.9	20.9	20.1	19.4	18.7	18.1	17.6	17.1	16.6	16.2	14.5	13.2	11.5	10.3	80
	90	38.4	34.4	31.4	29.0	27.2	25.6	24.3	23.2	22.2	21.3	20.5	19.8	19.2	18.6	18.1	17.6	17.2	15.4	14.0	12.2	10.9	90
	100	40.5	36.2	33.1	30.6	28.6	27.0	25.6	24.4	23.4	22.5	21.6	20.9	20.2	19.6	19.1	18.6	18.1	16.2	14.8	12.8	11.5	100

	0.42	4	5	6	7	8	9	10	11	12	13	14	15	16	17	18	19	20	25	30	40	50	1.80
	8 inches past hole																						
	3	7.4	6.6	6.0	5.6	5.2	4.9	4.7	4.4	4.2	4.1	3.9	3.8	3.7	3.6	3.5	3.4	3.3	2.9	2.7	2.3	2.1	3
	6	9.9	8.9	8.1	7.5	7.0	6.6	6.3	6.0	5.7	5.5	5.3	5.1	5.0	4.8	4.7	4.6	4.4	4.0	3.6	3.1	2.8	6
	9	11.9	10.7	9.8	9.0	8.4	8.0	7.6	7.2	6.9	6.6	6.4	6.2	6.0	5.8	5.6	5.5	5.3	4.8	4.4	3.8	3.4	9
	12	13.7	12.2	11.2	10.3	9.7	9.1	8.7	8.2	7.9	7.6	7.3	7.1	6.8	6.6	6.4	6.3	6.1	5.5	5.0	4.3	3.9	12
	15	15.2	13.6	12.4	11.5	10.8	10.1	9.6	9.2	8.8	8.4	8.1	7.9	7.6	7.4	7.2	7.0	6.8	6.1	5.6	4.8	4.3	15
	18	16.6	14.9	13.6	12.6	11.7	11.1	10.5	10.0	9.6	9.2	8.9	8.6	8.3	8.1	7.8	7.6	7.4	6.6	6.1	5.3	4.7	18
	21	17.9	16.0	14.6	13.5	12.7	11.9	11.3	10.8	10.3	9.9	9.6	9.2	8.9	8.7	8.4	8.2	8.0	7.2	6.5	5.7	5.1	21
	24	19.1	17.1	15.6	14.4	13.5	12.7	12.1	11.5	11.0	10.6	10.2	9.9	9.5	9.3	9.0	8.8	8.5	7.6	7.0	6.0	5.4	24
	27	20.2	18.1	16.5	15.3	14.3	13.5	12.8	12.2	11.7	11.2	10.8	10.4	10.1	9.8	9.5	9.3	9.0	8.1	7.4	6.4	5.7	27
	30	21.3	19.0	17.4	16.1	15.0	14.2	13.5	12.8	12.3	11.8	11.4	11.0	10.6	10.3	10.0	9.8	9.5	8.5	7.8	6.7	6.0	30
	40	24.5	21.9	20.0	18.5	17.3	16.3	15.5	14.8	14.2	13.6	13.1	12.7	12.3	11.9	11.6	11.2	11.0	9.8	8.9	7.8	6.9	40
	50	27.4	24.5	22.3	20.7	19.3	18.2	17.3	16.5	15.8	15.2	14.6	14.1	13.7	13.3	12.9	12.6	12.2	10.9	10.0	8.7	7.7	50
	60	29.9	26.8	24.4	22.6	21.2	20.0	18.9	18.1	17.3	16.6	16.0	15.5	15.0	14.5	14.1	13.7	13.4	12.0	10.9	9.5	8.5	60
	70	32.3	28.9	26.4	24.4	22.8	21.5	20.4	19.5	18.7	17.9	17.3	16.7	16.2	15.7	15.2	14.8	14.4	12.9	11.8	10.2	9.1	70
	80	34.5	30.9	28.2	26.1	24.4	23.0	21.8	20.8	19.9	19.1	18.5	17.8	17.3	16.7	16.3	15.8	15.4	13.8	12.6	10.9	9.8	80
	90	36.6	32.7	29.9	27.7	25.9	24.4	23.1	22.1	21.1	20.3	19.6	18.9	18.3	17.8	17.3	16.8	16.4	14.6	13.4	11.6	10.4	90
	100	38.6	34.5	31.5	29.1	27.3	25.7	24.4	23.3	22.3	21.4	20.6	19.9	19.3	18.7	18.2	17.7	17.2	15.4	14.1	12.2	10.9	100

1.85 Impact Ratio

Pmph/BS"

0.32	4	5	6	7	8	9	10	11	12	13	14	15	16	17	18	19	20	25	30	40	50	1.85
	8 inches past hole																					
3	9.4	8.4	7.7	7.1	6.6	6.3	5.9	5.7	5.4	5.2	5.0	4.9	4.7	4.6	4.4	4.3	4.2	3.8	3.4	3.0	2.7	3
6	12.7	11.3	10.3	9.6	9.0	8.4	8.0	7.6	7.3	7.0	6.8	6.5	6.3	6.1	6.0	5.8	5.7	5.1	4.6	4.0	3.6	6
9	15.3	13.6	12.5	11.5	10.8	10.2	9.7	9.2	8.8	8.5	8.2	7.9	7.6	7.4	7.2	7.0	6.8	6.1	5.6	4.8	4.3	9
12	17.5	15.6	14.3	13.2	12.4	11.6	11.0	10.5	10.1	9.7	9.3	9.0	8.7	8.5	8.2	8.0	7.8	7.0	6.4	5.5	4.9	12
15	19.4	17.4	15.9	14.7	13.7	13.0	12.3	11.7	11.2	10.8	10.4	10.0	9.7	9.4	9.2	8.9	8.7	7.8	7.1	6.1	5.5	15
18	21.2	19.0	17.3	16.0	15.0	14.1	13.4	12.8	12.2	11.8	11.3	11.0	10.6	10.3	10.0	9.7	9.5	8.5	7.7	6.7	6.0	18
21	22.8	20.4	18.7	17.3	16.2	15.2	14.4	13.8	13.2	12.7	12.2	11.8	11.4	11.1	10.8	10.5	10.2	9.1	8.3	7.2	6.5	21
24	24.4	21.8	19.9	18.4	17.2	16.3	15.4	14.7	14.1	13.5	13.0	12.6	12.2	11.8	11.5	11.2	10.9	9.8	8.9	7.7	6.9	24
27	25.8	23.1	21.1	19.5	18.3	17.2	16.3	15.6	14.9	14.3	13.8	13.3	12.9	12.5	12.2	11.8	11.5	10.3	9.4	8.2	7.3	27
30	27.2	24.3	22.2	20.5	19.2	18.1	17.2	16.4	15.7	15.1	14.5	14.0	13.6	13.2	12.8	12.5	12.2	10.9	9.9	8.6	7.7	30
40	31.3	28.0	25.6	23.7	22.1	20.9	19.8	18.9	18.1	17.4	16.7	16.2	15.6	15.2	14.8	14.4	14.0	12.5	11.4	9.9	8.9	40
50	34.9	31.2	28.5	26.4	24.7	23.3	22.1	21.1	20.2	19.4	18.7	18.0	17.5	16.9	16.5	16.0	15.6	14.0	12.8	11.0	9.9	50
60	38.2	34.2	31.2	28.9	27.0	25.5	24.2	23.1	22.1	21.2	20.4	19.7	19.1	18.5	18.0	17.5	17.1	15.3	14.0	12.1	10.8	60
70	41.3	36.9	33.7	31.2	29.2	27.5	26.1	24.9	23.8	22.9	22.1	21.3	20.6	20.0	19.4	18.9	18.5	16.5	15.1	13.0	11.7	70
80	44.1	39.4	36.0	33.3	31.2	29.4	27.9	26.6	25.5	24.5	23.6	22.8	22.0	21.4	20.8	20.2	19.7	17.6	16.1	13.9	12.5	80
90	46.7	41.8	38.2	35.3	33.0	31.2	29.6	28.2	27.0	25.9	25.0	24.1	23.4	22.7	22.0	21.4	20.9	18.7	17.1	14.8	13.2	90
100	49.2	44.0	40.2	37.2	34.8	32.8	31.1	29.7	28.4	27.3	26.3	25.4	24.6	23.9	23.2	22.6	22.0	19.7	18.0	15.6	13.9	100

0.34	4	5	6	7	8	9	10	11	12	13	14	15	16	17	18	19	20	25	30	40	50	1.85
	8 inches past hole																					
3	8.8	7.9	7.2	6.7	6.3	5.9	5.6	5.3	5.1	4.9	4.7	4.6	4.4	4.3	4.2	4.1	4.0	3.5	3.2	2.8	2.5	3
6	11.9	10.7	9.7	9.0	8.4	8.0	7.5	7.2	6.9	6.6	6.4	6.2	6.0	5.8	5.6	5.5	5.3	4.8	4.4	3.8	3.4	6
9	14.4	12.8	11.7	10.9	10.2	9.6	9.1	8.7	8.3	8.0	7.7	7.4	7.2	7.0	6.8	6.6	6.4	5.7	5.2	4.5	4.1	9
12	16.4	14.7	13.4	12.4	11.6	11.0	10.4	9.9	9.5	9.1	8.8	8.5	8.2	8.0	7.8	7.5	7.4	6.6	6.0	5.2	4.7	12
15	18.3	16.4	14.9	13.8	12.9	12.2	11.6	11.0	10.6	10.1	9.8	9.4	9.1	8.9	8.6	8.4	8.2	7.3	6.7	5.8	5.2	15
18	20.0	17.9	16.3	15.1	14.1	13.3	12.6	12.0	11.5	11.1	10.7	10.3	10.0	9.7	9.4	9.2	8.9	8.0	7.3	6.3	5.6	18
21	21.5	19.2	17.6	16.3	15.2	14.3	13.6	13.0	12.4	11.9	11.5	11.1	10.8	10.4	10.1	9.9	9.6	8.6	7.9	6.8	6.1	21
24	22.9	20.5	18.7	17.3	16.2	15.3	14.5	13.8	13.2	12.7	12.3	11.8	11.5	11.1	10.8	10.5	10.3	9.2	8.4	7.3	6.5	24
27	24.3	21.7	19.8	18.4	17.2	16.2	15.4	14.7	14.0	13.5	13.0	12.5	12.1	11.8	11.5	11.1	10.9	9.7	8.9	7.7	6.9	27
30	25.6	22.9	20.9	19.3	18.1	17.1	16.2	15.4	14.8	14.2	13.7	13.2	12.8	12.4	12.1	11.7	11.4	10.2	9.3	8.1	7.2	30
40	29.5	26.3	24.1	22.3	20.8	19.6	18.6	17.8	17.0	16.3	15.7	15.2	14.7	14.3	13.9	13.5	13.2	11.8	10.8	9.3	8.3	40
50	32.9	29.4	26.8	24.9	23.3	21.9	20.8	19.8	19.0	18.2	17.6	17.0	16.4	15.9	15.5	15.1	14.7	13.2	12.0	10.4	9.3	50
60	36.0	32.2	29.4	27.2	25.4	24.0	22.8	21.7	20.8	20.0	19.2	18.6	18.0	17.5	17.0	16.5	16.1	14.4	13.1	11.4	10.2	60
70	38.8	34.7	31.7	29.4	27.5	25.9	24.6	23.4	22.4	21.5	20.8	20.1	19.4	18.8	18.3	17.8	17.4	15.5	14.2	12.3	11.0	70
80	41.5	37.1	33.9	31.4	29.3	27.7	26.2	25.0	24.0	23.0	22.2	21.4	20.7	20.1	19.6	19.0	18.6	16.6	15.1	13.1	11.7	80
90	44.0	39.3	35.9	33.2	31.1	29.3	27.8	26.5	25.4	24.4	23.5	22.7	22.0	21.3	20.7	20.2	19.7	17.6	16.1	13.9	12.4	90
100	46.3	41.5	37.8	35.0	32.8	30.9	29.3	27.9	26.8	25.7	24.8	23.9	23.2	22.5	21.8	21.3	20.7	18.5	16.9	14.7	13.1	100

0.36	4	5	6	7	8	9	10	11	12	13	14	15	16	17	18	19	20	25	30	40	50	1.85
	8 inches past hole																					
3	8.4	7.5	6.8	6.3	5.9	5.6	5.3	5.0	4.8	4.6	4.5	4.3	4.2	4.1	3.9	3.8	3.7	3.3	3.1	2.6	2.4	3
6	11.3	10.1	9.2	8.5	8.0	7.5	7.1	6.8	6.5	6.2	6.0	5.8	5.6	5.5	5.3	5.2	5.0	4.5	4.1	3.6	3.2	6
9	13.6	12.1	11.1	10.3	9.6	9.0	8.6	8.2	7.8	7.5	7.3	7.0	6.8	6.6	6.4	6.2	6.1	5.4	5.0	4.3	3.8	9
12	15.5	13.9	12.7	11.7	11.0	10.4	9.8	9.4	9.0	8.6	8.3	8.0	7.8	7.5	7.3	7.1	6.9	6.2	5.7	4.9	4.4	12
15	17.3	15.4	14.1	13.1	12.2	11.5	10.9	10.4	10.0	9.6	9.2	8.9	8.6	8.4	8.1	7.9	7.7	6.9	6.3	5.5	4.9	15
18	18.8	16.9	15.4	14.2	13.3	12.6	11.9	11.4	10.9	10.5	10.1	9.7	9.4	9.1	8.9	8.6	8.4	7.5	6.9	6.0	5.3	18
21	20.3	18.2	16.6	15.4	14.4	13.5	12.8	12.2	11.7	11.3	10.9	10.5	10.2	9.9	9.6	9.3	9.1	8.1	7.4	6.4	5.7	21
24	21.7	19.4	17.7	16.4	15.3	14.4	13.7	13.1	12.5	12.0	11.6	11.2	10.8	10.5	10.2	9.9	9.7	8.7	7.9	6.9	6.1	24
27	22.9	20.5	18.7	17.3	16.2	15.3	14.5	13.8	13.2	12.7	12.3	11.9	11.5	11.1	10.8	10.5	10.3	9.2	8.4	7.3	6.5	27
30	24.2	21.6	19.7	18.3	17.1	16.1	15.3	14.6	13.9	13.4	12.9	12.5	12.1	11.7	11.4	11.1	10.8	9.7	8.8	7.6	6.8	30
40	27.8	24.9	22.7	21.0	19.7	18.5	17.6	16.8	16.1	15.4	14.9	14.4	13.9	13.5	13.1	12.8	12.4	11.1	10.2	8.8	7.9	40
50	31.1	27.8	25.4	23.5	22.0	20.7	19.6	18.7	17.9	17.2	16.6	16.0	15.5	15.1	14.6	14.2	13.9	12.4	11.3	9.8	8.8	50
60	34.0	30.4	27.7	25.7	24.0	22.7	21.5	20.5	19.6	18.8	18.2	17.5	17.0	16.5	16.0	15.6	15.2	13.6	12.4	10.7	9.6	60
70	36.7	32.8	29.9	27.7	25.9	24.4	23.2	22.1	21.2	20.3	19.6	18.9	18.3	17.8	17.3	16.8	16.4	14.7	13.4	11.6	10.4	70
80	39.2	35.0	32.0	29.6	27.7	26.1	24.8	23.6	22.6	21.7	20.9	20.2	19.6	19.0	18.5	18.0	17.5	15.7	14.3	12.4	11.1	80
90	41.5	37.2	33.9	31.4	29.4	27.7	26.3	25.1	24.0	23.0	22.2	21.5	20.8	20.2	19.6	19.1	18.6	16.6	15.2	13.1	11.7	90
100	43.8	39.2	35.7	33.1	31.0	29.2	27.7	26.4	25.3	24.3	23.4	22.6	21.9	21.2	20.6	20.1	19.6	17.5	16.0	13.8	12.4	100

1.85 Impact Ratio

Pmph/BS"

0.38	4	5	6	7	8	9	10	11	12	13	14	15	16	17	18	19	20	25	30	40	50	1.85
8 inches past hole																						Impact Ratio
3	7.9	7.1	6.5	6.0	5.6	5.3	5.0	4.8	4.6	4.4	4.2	4.1	4.0	3.8	3.7	3.6	3.5	3.2	2.9	2.5	2.2	3
6	10.7	9.5	8.7	8.1	7.5	7.1	6.7	6.4	6.2	5.9	5.7	5.5	5.3	5.2	5.0	4.9	4.8	4.3	3.9	3.4	3.0	6
9	12.9	11.5	10.5	9.7	9.1	8.6	8.1	7.7	7.4	7.1	6.9	6.6	6.4	6.2	6.1	5.9	5.7	5.1	4.7	4.1	3.6	9
12	14.7	13.2	12.0	11.1	10.4	9.8	9.3	8.9	8.5	8.2	7.9	7.6	7.4	7.1	6.9	6.7	6.6	5.9	5.4	4.7	4.2	12
15	16.4	14.6	13.4	12.4	11.6	10.9	10.3	9.9	9.4	9.1	8.7	8.4	8.2	7.9	7.7	7.5	7.3	6.5	6.0	5.2	4.6	15
18	17.9	16.0	14.6	13.5	12.6	11.9	11.3	10.8	10.3	9.9	9.5	9.2	8.9	8.7	8.4	8.2	8.0	7.1	6.5	5.6	5.1	18
21	19.2	17.2	15.7	14.5	13.6	12.8	12.2	11.6	11.1	10.7	10.3	9.9	9.6	9.3	9.1	8.8	8.6	7.7	7.0	6.1	5.4	21
24	20.5	18.4	16.8	15.5	14.5	13.7	13.0	12.4	11.9	11.4	11.0	10.6	10.3	10.0	9.7	9.4	9.2	8.2	7.5	6.5	5.8	24
27	21.7	19.4	17.8	16.4	15.4	14.5	13.7	13.1	12.6	12.1	11.6	11.2	10.9	10.5	10.2	10.0	9.7	8.7	7.9	6.9	6.1	27
30	22.9	20.5	18.7	17.3	16.2	15.3	14.5	13.8	13.2	12.7	12.2	11.8	11.4	11.1	10.8	10.5	10.2	9.2	8.4	7.2	6.5	30
40	26.4	23.6	21.5	19.9	18.6	17.6	16.7	15.9	15.2	14.6	14.1	13.6	13.2	12.8	12.4	12.1	11.8	10.5	9.6	8.3	7.5	40
50	29.4	26.3	24.0	22.2	20.8	19.6	18.6	17.7	17.0	16.3	15.7	15.2	14.7	14.3	13.9	13.5	13.2	11.8	10.7	9.3	8.3	50
60	32.2	28.8	26.3	24.3	22.8	21.5	20.4	19.4	18.6	17.9	17.2	16.6	16.1	15.6	15.2	14.8	14.4	12.9	11.8	10.2	9.1	60
70	34.7	31.1	28.4	26.3	24.6	23.2	22.0	21.0	20.1	19.3	18.6	17.9	17.4	16.9	16.4	15.9	15.5	13.9	12.7	11.0	9.8	70
80	37.1	33.2	30.3	28.1	26.2	24.7	23.5	22.4	21.4	20.6	19.8	19.2	18.6	18.0	17.5	17.0	16.6	14.8	13.6	11.7	10.5	80
90	39.4	35.2	32.1	29.7	27.8	26.2	24.9	23.7	22.7	21.8	21.0	20.3	19.7	19.1	18.6	18.1	17.6	15.7	14.4	12.4	11.1	90
100	41.5	37.1	33.9	31.3	29.3	27.6	26.2	25.0	23.9	23.0	22.2	21.4	20.7	20.1	19.5	19.0	18.5	16.6	15.1	13.1	11.7	100

0.40	4	5	6	7	8	9	10	11	12	13	14	15	16	17	18	19	20	25	30	40	50	1.85
8 inches past hole																						
3	7.5	6.7	6.1	5.7	5.3	5.0	4.8	4.5	4.3	4.2	4.0	3.9	3.8	3.6	3.5	3.4	3.4	3.0	2.7	2.4	2.1	3
6	10.1	9.1	8.3	7.7	7.2	6.8	6.4	6.1	5.9	5.6	5.4	5.2	5.1	4.9	4.8	4.7	4.5	4.1	3.7	3.2	2.9	6
9	12.2	10.9	10.0	9.2	8.6	8.1	7.7	7.4	7.0	6.8	6.5	6.3	6.1	5.9	5.8	5.6	5.5	4.9	4.5	3.9	3.5	9
12	14.0	12.5	11.4	10.6	9.9	9.3	8.8	8.4	8.1	7.8	7.5	7.2	7.0	6.8	6.6	6.4	6.2	5.6	5.1	4.4	4.0	12
15	15.5	13.9	12.7	11.7	11.0	10.4	9.8	9.4	9.0	8.6	8.3	8.0	7.8	7.5	7.3	7.1	7.0	6.2	5.7	4.9	4.4	15
18	17.0	15.2	13.9	12.8	12.0	11.3	10.7	10.2	9.8	9.4	9.1	8.8	8.5	8.2	8.0	7.8	7.6	6.8	6.2	5.4	4.8	18
21	18.3	16.3	14.9	13.8	12.9	12.2	11.6	11.0	10.6	10.1	9.8	9.4	9.1	8.9	8.6	8.4	8.2	7.3	6.7	5.8	5.2	21
24	19.5	17.4	15.9	14.7	13.8	13.0	12.3	11.8	11.3	10.8	10.4	10.1	9.8	9.5	9.2	8.9	8.7	7.8	7.1	6.2	5.5	24
27	20.7	18.5	16.9	15.6	14.6	13.8	13.1	12.5	11.9	11.5	11.0	10.7	10.3	10.0	9.7	9.5	9.2	8.3	7.5	6.5	5.8	27
30	21.7	19.4	17.8	16.4	15.4	14.5	13.8	13.1	12.6	12.1	11.6	11.2	10.9	10.5	10.3	10.0	9.7	8.7	7.9	6.9	6.2	30
40	25.0	22.4	20.4	18.9	17.7	16.7	15.8	15.1	14.5	13.9	13.4	12.9	12.5	12.1	11.8	11.5	11.2	10.0	9.1	7.9	7.1	40
50	27.9	25.0	22.8	21.1	19.8	18.6	17.7	16.9	16.1	15.5	14.9	14.4	14.0	13.6	13.2	12.8	12.5	11.2	10.2	8.8	7.9	50
60	30.6	27.4	25.0	23.1	21.6	20.4	19.3	18.4	17.7	17.0	16.3	15.8	15.3	14.8	14.4	14.0	13.7	12.2	11.2	9.7	8.7	60
70	33.0	29.5	27.0	25.0	23.3	22.0	20.9	19.9	19.1	18.3	17.6	17.0	16.5	16.0	15.6	15.1	14.8	13.2	12.1	10.4	9.3	70
80	35.3	31.5	28.8	26.7	24.9	23.5	22.3	21.3	20.4	19.6	18.9	18.2	17.6	17.1	16.6	16.2	15.8	14.1	12.9	11.2	10.0	80
90	37.4	33.4	30.5	28.3	26.4	24.9	23.6	22.5	21.6	20.7	20.0	19.3	18.7	18.1	17.6	17.2	16.7	15.0	13.7	11.8	10.6	90
100	39.4	35.2	32.2	29.8	27.9	26.3	24.9	23.8	22.7	21.9	21.1	20.3	19.7	19.1	18.6	18.1	17.6	15.8	14.4	12.5	11.1	100

0.42	4	5	6	7	8	9	10	11	12	13	14	15	16	17	18	19	20	25	30	40	50	1.85
8 inches past hole																						
3	7.2	6.4	5.8	5.4	5.1	4.8	4.5	4.3	4.1	4.0	3.8	3.7	3.6	3.5	3.4	3.3	3.2	2.9	2.6	2.3	2.0	3
6	9.7	8.6	7.9	7.3	6.8	6.4	6.1	5.8	5.6	5.4	5.2	5.0	4.8	4.7	4.6	4.4	4.3	3.9	3.5	3.1	2.7	6
9	11.6	10.4	9.5	8.8	8.2	7.8	7.4	7.0	6.7	6.4	6.2	6.0	5.8	5.6	5.5	5.3	5.2	4.7	4.2	3.7	3.3	9
12	13.3	11.9	10.9	10.1	9.4	8.9	8.4	8.0	7.7	7.4	7.1	6.9	6.7	6.5	6.3	6.1	6.0	5.3	4.9	4.2	3.8	12
15	14.8	13.2	12.1	11.2	10.5	9.9	9.4	8.9	8.5	8.2	7.9	7.6	7.4	7.2	7.0	6.8	6.6	5.9	5.4	4.7	4.2	15
18	16.2	14.5	13.2	12.2	11.4	10.8	10.2	9.7	9.3	9.0	8.6	8.3	8.1	7.8	7.6	7.4	7.2	6.5	5.9	5.1	4.6	18
21	17.4	15.6	14.2	13.2	12.3	11.6	11.0	10.5	10.0	9.7	9.3	9.0	8.7	8.4	8.2	8.0	7.8	7.0	6.4	5.5	4.9	21
24	18.6	16.6	15.2	14.0	13.1	12.4	11.7	11.2	10.7	10.3	9.9	9.6	9.3	9.0	8.8	8.5	8.3	7.4	6.8	5.9	5.3	24
27	19.7	17.6	16.1	14.9	13.9	13.1	12.4	11.9	11.4	10.9	10.5	10.2	9.8	9.5	9.3	9.0	8.8	7.9	7.2	6.2	5.6	27
30	20.7	18.5	16.9	15.7	14.6	13.8	13.1	12.5	12.0	11.5	11.1	10.7	10.4	10.0	9.8	9.5	9.3	8.3	7.6	6.5	5.9	30
40	23.8	21.3	19.5	18.0	16.9	15.9	15.1	14.4	13.8	13.2	12.7	12.3	11.9	11.6	11.2	10.9	10.7	9.5	8.7	7.5	6.7	40
50	26.6	23.8	21.7	20.1	18.8	17.7	16.8	16.1	15.4	14.8	14.2	13.7	13.3	12.9	12.5	12.2	11.9	10.6	9.7	8.4	7.5	50
60	29.1	26.1	23.8	22.0	20.6	19.4	18.4	17.6	16.8	16.2	15.6	15.0	14.6	14.1	13.7	13.4	13.0	11.7	10.6	9.2	8.2	60
70	31.4	28.1	25.7	23.8	22.2	21.0	19.9	19.0	18.1	17.4	16.8	16.2	15.7	15.2	14.8	14.4	14.1	12.6	11.5	9.9	8.9	70
80	33.6	30.0	27.4	25.4	23.7	22.4	21.2	20.3	19.4	18.6	18.0	17.3	16.8	16.3	15.8	15.4	15.0	13.4	12.3	10.6	9.5	80
90	35.6	31.8	29.1	26.9	25.2	23.7	22.5	21.5	20.6	19.8	19.0	18.4	17.8	17.3	16.8	16.3	15.9	14.2	13.0	11.3	10.1	90
100	37.5	33.6	30.6	28.4	26.5	25.0	23.7	22.6	21.7	20.8	20.1	19.4	18.8	18.2	17.7	17.2	16.8	15.0	13.7	11.9	10.6	100

Appendix B
Formulas

Pendulum Physics Formulas

The period of a pendulum (t) varies with pendulum length (L) and the constant acceleration (g) force:

$$t = 2\pi\sqrt{L/g} \qquad (1.1)$$

When gravity is the constant acceleration force (g = 9.81 m/sec²) period varies only with length (L).
Solving for the putting stroke constant acceleration (g_{alt}):

$$g_{alt} = L/[t^2/((2\pi)^2)] \qquad\qquad g_{alt} = \frac{4\pi^2 L}{t^2} \qquad (1.2)$$

The velocity of the putter at impact (v_{max}) varies with pendulum length, constant acceleration force (g) and amplitude of the pendulum (θ). The amplitude is the angle of the putter shaft from vertical at the back of the backstroke.

$$v_{max} = \sqrt{2gL(1-\cos\theta)} \qquad (1.3)$$

$$v_{max} = \sqrt{2g_{alt}L(1-\cos\theta)} \qquad (1.4)$$

The v_{max} required for a putt of a specific length is determined by the backstroke length with a regularized pendulum period of from 1.1 to 1.3 seconds (t), and a fixed pendulum length (L):

$$\cos\theta = 1 - \frac{v_{max}^2}{2gL} \qquad (1.5)$$

Force (f) from gravity acting on the putter head (pendulum bob):

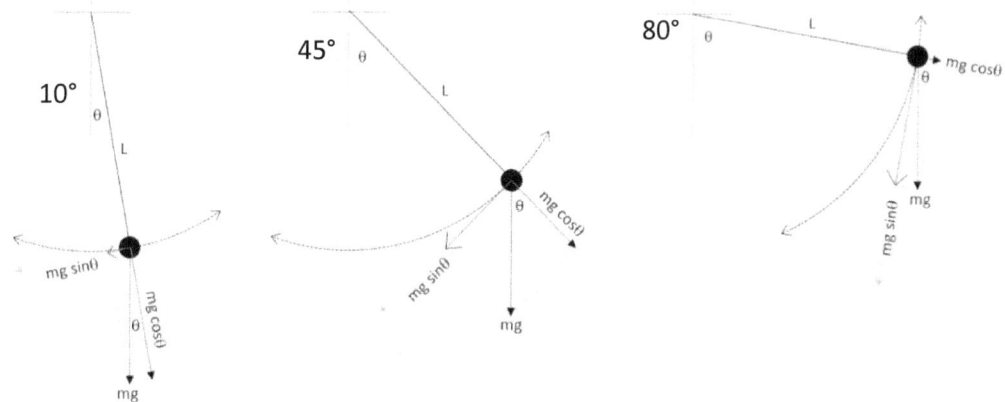

mg (mass x gravity) is constant: e.g. 0.375 kg x 9.81 m/sec² = 3.68 Newtons.
The vector of the force (f) acting on the bob increases dramatically with amplitude.

Kinetic energy (K) varies with mass (m) and velocity (v) squared:

$$K = \frac{1}{2}mv^2 \qquad (1.8)$$

Coefficient of Rolling Friction Formulas

Friction force (F_f) = μ times mass times gravity:

$$F_f = \mu_r mg \qquad (3.1)$$

Palmer, Physics p186

Coefficient of rolling friction (μ_r):

$$\mu_r = \frac{F_r}{N} \qquad (3.2)$$

Ron Kurtus

Coefficient of rolling friction (μ_r) of a golf ball rolling on a putting green:

$$\mu_r = \frac{7}{5}\left(\frac{0.48}{stimp}\right) \qquad (3.3)$$

Roh/Lee from Penner, Physics

$$stimp = \frac{0.672}{\mu_r} \qquad (3.4)$$

Roh/Lee from Penner, Physics

Putt Distance Formulas

The Impact Phase

The impact flight distance is the small flight distance resulting from impact with a non-zero loft putter. Because the launch angle from loft of the putter is very small, Palmer (Physics p188) assumes that the post collision velocity of the ball will be the same as the post-collision normal velocity.

The line of action will be related to the loft of the putter and launch angle of the ball. Palmer (Physics p188) determines that calculating this initial ball flight is so small, and inconsequential, that it is not worth the trouble of calculating it. Assuming the initial flight distance to be on the order of 1.5% of putt distance. The "hop" ranges on a 2.99 m putt from 0.04 m (1.34% of putt length) to 0.08 m (2.65% of putt length). An average of 0.06 m (2.00% of putt length). Palmer assumes a fixed 0.06 m for ALL putts.

$$x_i = 0.04\ m = impact\ flight\ distance = 1.57\ inches$$

$$x_i = 0.06\ m = impact\ flight\ distance = 2.36\ inches$$

$$x_i = 0.08\ m = impact\ flight\ distance = 3.14\ inches$$

Palmer, Physics p185

SRD/MJD tentative formula
Impact Flight Distance = 0.15 times impact ball velocity squared

$$x_i = 0.15v^2 \qquad (4.1)$$

SRD/MJD

Werner & Greig provide a slightly more complicated formula factoring in launch angle (α)

$$x_i = [(2v_i^2)(\cos \alpha)(\sin \alpha)]/g \qquad (4.2)$$

Werner & Greig, How p127

The Werner & Greig formula yields shorter impact flight distances with x_i = about 1.0" on a 10' putt with 2° launch angle; and ranging to zero with zero launch angle. There are very complicated formulas for impact flight distance but the additional precision is not needed. Following Palmer, assume 0.04 m (1.57 inches) for all putts.

Quintic recommends 0.75° to 2.0° launch angle for most greens (lower launch angle with faster greens). Specifically recommending against lower the 0.75° launch angle, to get the ball out of its nest. Michael Breed research indicates that zero launch angle or even up to 3° negative launch angle produce better results.

Putt Distance Formulas

The Skid Phase

After impact and the initial small hop, the ball skids along the green until true roll is achieved. The ball starts in a condition of almost pure skid, the absence of rolling. In this condition the lower surface of the ball is traveling at approximately the same speed as the center of the ball. The percentage of skid distance (x_s) is approximately constant irrespective of total putt distance (x_{putt}).

$$x_s = \frac{1}{7}x_{putt} \qquad (5.1)$$

Dewhurst, Science p202

The coefficient of sliding friction (μ_s) will range from 0.4 to 0.5

Weber, Physics
Palmer, Physics p187

$$x_s = \frac{1}{6}x_r \qquad (5.2)$$

$$x_s + x_r = \frac{7}{6}x_r \qquad (5.3)$$

Putt Distance Formulas

The Roll Phase

Velocity of ball at beginning of roll is 5/7 of ball velocity at the beginning of skid

$$v_{br} = \frac{5}{7}(v_p \times 1.70) \tag{6.1}$$

$$v_{br} = \frac{5}{7} v_{bs} \tag{6.2}$$

After skid, the distance a ball will roll (x) can be calculated by the following formula, based on velocity at the beginning of roll (v), coefficient of rolling friction (μ) and gravity (g).

$$x = \frac{v^2}{2\mu g} \tag{6.3}$$

After skid, the distance a ball will roll (x_r) can be calculated by the following formula, based on velocity at the beginning of roll (v_{br}), coefficient of rolling friction (μ_r) and gravity (g).

$$x_r = \frac{v_{br}^2}{2\mu_r g} \tag{6.4}$$

Palmer, Physics p202

Solving for µ

$$\mu = \frac{v^2}{2gx} \tag{6.5}$$

Palmer, Physics p202

$$\mu_r = \frac{v_{br}^2}{2gx_r} \tag{6.6}$$

Palmer, Physics p202

Putt Distance Formulas

Total Putt Length

$$x_i = 0.04 \, m = impact \, flight \, distance = 1.57 \, inches$$

$$x_s = \frac{x_{putt}}{7} \tag{7.1}$$

$$x_r = \frac{v_{br}^2}{2(\mu_r)(g)} \tag{7.2}$$

$$x_r + x_s = \frac{7}{6}\left(\frac{v_{br}^2}{2(\mu_r)(g)}\right) \tag{7.3}$$

$$x_r + x_s + x_i = 0.04 + \frac{7}{6}\left(\frac{v_{br}^2}{2(\mu_r)(g)}\right) \tag{7.4}$$

$$x_{putt} = 0.04 + \frac{7}{6}\left(\frac{v_{br}^2}{2(\mu_r)(g)}\right) \tag{7.5}$$

Misc additional formulas:

$$g_{alt} = (v_{max}^2)/2L(1 - \cos\theta) \tag{8.1}$$

www.ingramcontent.com/pod-product-compliance
Lightning Source LLC
Chambersburg PA
CBHW081200230426
43666CB00016B/2872